04/22

EXTINCTION and RADIATION

EXTINCTION and RADIATION

How the Fall of Dinosaurs Led to the Rise of Mammals

J. DAVID ARCHIBALD

THE JOHNS HOPKINS UNIVERSITY PRESS

Baltimore

© 2011 The Johns Hopkins University Press
All rights reserved. Published 2011
Printed in the United States of America on acid-free paper
9 8 7 6 5 4 3 2 1

The Johns Hopkins University Press
2715 North Charles Street
Baltimore, Maryland 21218-4363
www.press.jhu.edu

Library of Congress Cataloging-in-Publication Data

Archibald, J. David.
    Extinction and radiation : how the fall of dinosaurs led to the rise of
mammals / J. David Archibald.
        p. cm.
    Includes bibliographical references and index.
    ISBN-13: 978-0-8018-9805-1 (hardcover : alk. paper)
    ISBN-10: 0-8018-9805-6 (hardcover : alk. paper)
    1. Dinosaurs—Extinction. 2. Cretaceous-Tertiary boundary.
3. Mammals, Fossil. 4. Mammals. I. Title.
    QE861.6.E95A73 2011
    569—dc22   2010022281

A catalog record for this book is available from the British Library.

Title page illustration: Study of the basal eutherian *Kulbeckia.* Reconstruction by Maria Gonzalez.

*Special discounts are available for bulk purchases of this book. For
more information, please contact Special Sales at 410-516-6936 or
specialsales@press.jhu.edu.*

The Johns Hopkins University Press uses environmentally friendly
book materials, including recycled text paper that is composed of at
least 30 percent post-consumer waste, whenever possible. All of our
book papers are acid-free, and our jackets and covers are printed on
paper with recycled content.

For my best friend, Gloria Eugenia Bader

CONTENTS

PREFACE

WHILE WRITING THIS BOOK I was directed by my friend Milt Lessner to a cover story in *Newsweek* magazine. The story had the rather tabloid title "Who Was More Important: Lincoln or Darwin?"[1] Because I am an evolutionary biologist and a collector of Darwin, Milt knew the article might interest me. When I saw the cover, portraying both men, with the answer in small print declaring that Lincoln was the more important, my reaction was to roll my eyes and shake my head. The only real connection between these two men was that the bicentennial of their births was then fast approaching. In fact, both share the same date and year of birth: February 12, 1809.

I could not agree with the assessment of Lincoln's greater importance because Darwin's work shook the entire worldview, whereas Lincoln was instrumental in holding one country together and emancipating its slaves. No matter how important to the future of the United States, Lincoln's stage was basically one country, but Darwin influenced all humankind, beginning with Europe and North America but soon spreading elsewhere.

While the comparisons were somewhat strained, to my surprise I found the article quite good, and by the end I found myself agreeing with the author. Although both men were revolutionaries in their respective endeavors, Lincoln was unique for what he did in his time and place, while Darwin arguably was not. Darwin had planned a much longer volume than the one he rushed to completion in 1859, which comes down to us as *On the Origin of Species*. Darwin had been mulling over his ideas about the primary cause of evolution for almost 20 years, but this all changed when he received a letter in 1858 from the much younger Alfred Russel Wallace laying out the same cause of evolution: natural selection.

The author of the *Newsweek* piece did not take this argument to completion, but what he had demonstrated was the much more general idea that sets the natural sciences apart from all other endeavors. Discoveries are out there for the finding in natural science, but not so with other pursuits, such as art or politics. Artists do not discover a school of painting; they invent it. Politicians do not discover political movements; they create them. Inventors and engineers of course use the natural sciences all the time in their work, but they are applying scientific ideas, not doing science per se. This is the frequent confusion made between technology and science. They are not synonymous.

While I have oversimplified the case to some degree, I do strongly argue that, although scientists are just as much a product of their time and place as are other humans, discovery, not invention or creation, is the fundamental difference between science and other human pursuits. If Darwin had not discovered natural selection, the time was close for another brilliant scientist, such as Wallace, to bring this process to light. Of course, because of Darwin's stature the theory of evolution by means of natural selection received much more attention than if the lesser-known Wallace had been the sole author. By contrast, if there had been no Lincoln, the course of American history may well have been very different.

Beyond the simple fact that I am in general fascinated by what sets science apart from other pursuits, I am more specifically fascinated by the numerous and varied attempts to discover what happened on earth some 66 million years ago to change our planet forever. These discoveries must be pieced together carefully to yield coherent portraits of what transpired. I write *portraits* rather than *portrait* because I do not believe we can, as of now, give a definitive answer as to what transpired, or why. I do not know if we will ever have "the" answer, but I am confident that we will continue to narrow down the possibilities.

More and better fossils are being discovered; new and better sections will certainly come to light; even better methods—for molecular systematics, for example—will be created; and we will be able to better theorize about the physical consequences of what transpired so very long ago, when the ground-dwelling dinosaurs disappeared, to be supplanted by mammals. Our own human history is bound up in this drama, for as often has been noted, without dinosaur extinction, we would almost certainly not be here to discover, create, or invent anything.

I was bemused some years ago by an online reviewer who was upset that I did not provide the definitive answer to "what killed the dinosaurs" in my 1996 book on dinosaur extinction.[2] That is exactly the point. There is no single answer, or at least not one that we can provide with any certainty. I have shaken my head in disbelief more than once when some advocates of the single-cause impact theory of extinction at the K/T boundary exclaimed that, because an impact temporally correlated with the K/T boundary, it must be *the* cause of dinosaur extinction—an excellent example of the fallacy that correlation equals causation.

In a review I did of Doug Erwin's most recent book on the "mother of all extinctions" near the Permo-Triassic boundary,[3] I quoted his take on the issue of being too simplistic when it comes to complex events in the past. He said that "life would be much easier if complex events had single causes, but the lessons of history are otherwise."[4] He noted how we

try too often to find single causes for why civilizations fall.[5] If we cannot even determine with certainty why (or even when) the Roman Empire fell some 1,500 years ago,[6] how can we possibly be so certain about what caused a series of extinctions followed by radiations some 66 million years ago?

The book is organized into six chapters. The first chapter reviews the most recent data on what we know of dinosaur diversity decline during the last 10 million years of the Cretaceous in western North America and the still problematic issues of determining what was happening to dinosaur diversity just before the end of the Cretaceous. Chapters 2 and 3 deal with mammals. Chapter 2 introduces Mesozoic and especially Cretaceous mammals, explaining the history of our understanding of how these animals existed alongside dinosaurs. Chapter 3 goes over the difficult and sometimes thankless task of trying to find the ancestry of modern mammals within the ranks of Cretaceous mammals. The fourth and fifth chapters review what we currently know about how major groups of terrestrial organisms fared at the Cretaceous/Tertiary (K/T) boundary, notably in western North America, and what the best-supported hypotheses are concerning what caused extinctions at this boundary. The sixth and final chapter reflects on how the extinction of nonavian dinosaurs allowed the evolutionary diversification of mammals. It discusses where and how this happened and why there is a seeming disconnect between molecular and fossil evidence for this radiation.

I thank Daniel Archibald, Alexander Averianov, Bill Clemens, Tony Harper, Hans-Dieter Sues, David Ward, Greg Wilson, and anonymous Johns Hopkins University Press reviewers for reading and providing general comments, and Alexander Averianov, Bill Clemens, Tony Harper, and Greg Wilson for detailed comments on the text. The following kindly permitted use of their illustrative materials: John Alroy, the Geological Society of America, Marcelo Sanchez-Villagra, Henning Scholz, John Wible, and Greg Wilson. I prepared the figures, using original artwork by Maria Gonzalez and unpublished data graciously supplied by John Alroy, Alexander Averianov, Bill Clemens, Phil Currie, Mark Goodwin, and Greg Wilson. My thanks to my editor at the Johns Hopkins University Press, Vince Burke, for shepherding the book over various hurdles, and to Barbara Lamb, for copyediting the manuscript.

EXTINCTION and RADIATION

1

The Late Cretaceous Nonavian Dinosaur Record

Too OFTEN IN THE STUDY of the history of life on this planet an arbitrary demarcation is made in the continuum of geological time. We have books on the Cenozoic record of mammals, on trilobites of the Cambrian, and on the Mesozoic floras of Europe. There are exceptions, notably studies that deal with evolutionary relationships of organisms, but when there is an ecologic or, more correctly, paleoecologic component to the study, it often ends or starts at an obvious change in the fossil record. Nowhere is this better exemplified than in the study of biologic change across the clearly recognizable Cretaceous/Tertiary (K/T) boundary.[1] There are studies that deal with both sides of the boundary, but seldom do they deal with the biologic consequences as we pass through this boundary and literally enter a new geologic era—the Cenozoic. There are many books out there with the word *dinosaur* in the title that claim to have something to say about the extinction of these animals, but when examined, a surprising few do more than give dinosaurs proverbial lip service.

In this chapter I examine in detail one short interval of time in the geologic continuum, namely, the record of the decline and extinction of nonavian dinosaurs.[2] Important side issues come into play, such as the quality of the fossil record and how the quality of the record affects our perceptions of nonavian dinosaur decline and extinction. The final chapter concerns how the record colors our perception of when modern or extant mammals began to radiate.[3] These are controversial issues that require some space to explore.

WE ALL HAVE BIASES

Dinosaur fossils are not common, with a few notable exceptions. Even in the places and for the times that they are quite common, they still tend to be far rarer than most other fossil remains. The reasons for this rarity have nothing to do with dinosaurs per se. Rather, large animals such as nonavian dinosaurs tend to have smaller population sizes than smaller animals and thus have fewer potential remains that might end up in the fossil record. Further, as land animals, most nonavian dinosaurs tended to die in environments in which the chances of fossilization were quite low. Except for bodies of fresh water, or much rarer, shifting dunes, caves, or tar seeps, most of the land was an area of net erosion. This means that the land itself and anything that dies upon it is generally subject to destructive processes.

When we are looking for the fossils of land creatures, we generally look for lake or river deposits. If the conditions are right, lakes can preserve annual cycles of sedimentation, which may also record populations of such animals as fish. The Eocene Green River Formation of Wyoming preserves such records, not only of fish but also more rarely of bats and birds. This can also happen in marine rocks, such as the famous Jurassic Solenhofen Limestone, in Germany, which has produced fossils of the earliest bird, *Archaeopteryx*. What are needed for such preservation are fast burial and anoxic conditions, which preclude or minimize scavenging, or a combination of these factors.

These conditions are most dramatically seen following cataclysmic events, such as a violent volcanic eruption in which material borne from or set loose by the eruption can kill and bury many individuals—as happened with Vesuvius in 79 CE. If the eruption does not burn the corpse to a crisp, it can preserve skeletal remains and sometimes even an outline of the dead animal. Closer in time to the K/T boundary is the now famous Jehol paleofauna of China, dated from the Early Cretaceous (about 133 to 120 million years ago).[4] Here, fossils are often articulated skeletons showing remains of soft tissue and sometimes even stomach contents. Unlike Vesuvius, which was a series of cataclysmic eruptions over a few days, Jehol was produced in lakes punctuated by frequent volcanic ash falls. This helped to quickly bury organic remains before they rotted.

Even if some of these conditions were met in the geologic past and there are dinosaur fossils out there somewhere, they may not be for the taking because present geologic conditions must also be just right. The richest areas to find dinosaur fossils are appropriately known as badlands by the sane individuals who first saw them. But for paleontologists they are anything but "bad." These are usually dry areas, mostly deserts, stripped of much vegetation and actively being eroded by wind and water. Thus it comes as no surprise that the much drier areas of western North America have produced many more fossil vertebrates, including dinosaurs, than have the midwestern and eastern portions of the continent.

In addition to these natural biases affecting the fossil record of dinosaurs, there are human-caused biases as well. Regions and countries of the world that have been politically more stable and easier to access have generally yielded a greater bounty of nonavian dinosaurs. In the past twenty years we have seen a rapid expansion of paleontologic work on all continents, with spectacular results. Nonetheless, at the present time there is a clear bias for the fossil record of nonavian dinosaurs in the Western Interior of North America. Specifically for the Late Cretaceous record of nonavian dinosaurs, for all the reasons just discussed, North America has the best sequence of nonavian dinosaurs for this interval of time, bar none.

Even more germane for this discussion is that North America currently has the best and arguably the only vertebrate sequence that includes the K/T boundary. This means that not only is there a record of nonavian dinosaurs from beds near the end of the Cretaceous in western North America but there is also a record of other Late Cretaceous vertebrates. Finally, there is a record of vertebrates in the immediately overlying early Tertiary beds as well. As we will see, this has a profound affect on what we can and cannot say about the record of nonavian dinosaur extinction. It is unfortunate that western North America is at present the only place that offers fossiliferous terrestrial beds up to and through the K/T boundary. Whether this record is indicative of the rest of the terrestrial record around the world is a question only more fossiliferous geologic sections from elsewhere can answer.

While I understand the reluctance of some to use western North America as the surrogate for what we know of the terrestrial K/T boundary, some reactions have been rather strident. The French paleontologist Eric Buffetaut declared that to concentrate on a few counties in Montana while neglecting the rest of the world "shows a singular scientific myopia."[5] He suggested other places, such as Nanxiong Basin, southern China, but warned, as have others, that for the time being placement of the K/T there is problematic. He then offered parts of southern France and northern Spain as places "where it is possible to apprehend the events" at the K/T boundary.[6] While this small region of about 200 km on a side has produced some very nice Late Cretaceous nonavian dinosaurs, the K/T boundary there is really only now beginning to be investigated in any detail.[7] As noted earlier, such a boundary requires fossil-producing beds before, at, and after the boundary. These sites in Europe do not yet provide these sections. Most of the earliest part of the Paleocene (earliest Tertiary) terrestrial record in most of Europe is poorly known. By contrast, hundreds if not thousands of vertebrate localities surrounding *both* sides of the K/T boundary stretch some 1,200 km from southern Alberta south to eastern Wyoming, with some slightly less complete records dotted along a stretch all the way to northwestern New Mexico and southwestern Texas, an additional 1,500 km.

An interesting side issue for the sections in southwestern Europe is how they are interpreted regarding events at the K/T boundary. Buffetaut regards the fossil record there as supporting a catastrophic extinction, whereas Oms and his colleagues more recently stated that in Vallcebre Basin of the southeastern Spanish Pyrenees, during the last six million years of the Cretaceous through the K/T boundary, there was a gradual marine regression shifting from steady, low-energy sedimentation in better drained environments and culminating in higher energy deposition of sandstone. This sounds very much like conditions in the Western Interior of North America. Unlike Buffetaut's arguments for the K/T boundary in Europe, Oms and his colleagues say that "this scenario is consistent with the model for the dinosaur extinction proposed by Dingus and Rowe (1997), who suggested that a marine regression, in combination with latest Cretaceous igneous activity and the asteroid impact, caused the extinction."[8]

I had hoped that the K/T terrestrial record would have improved dramatically in the more than 12 years since I first

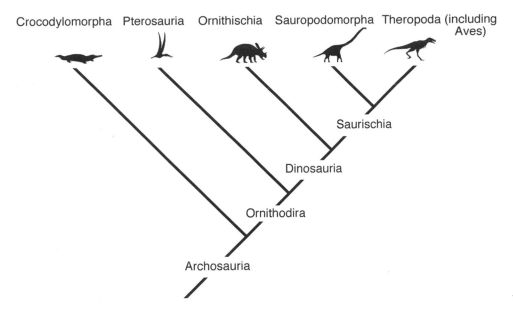

Crocodylomorpha Pterosauria Ornithischia Sauropodomorpha Theropoda (including Aves)

Saurischia

Dinosauria

Ornithodira

Archosauria

Fig. 1.1. Cladogram of Archosauria.

wrote a book on this topic, but unfortunately it has not. In the past, areas in Africa, South America, and China showed promise for good K/T sections, but to date only South America shows real promise. As for Europe, the areas in the Pyrenees show promise, but this has yet to be brought to fruition. For now, we are left with only western North America to serve as a proxy for the turnover of vertebrates at the K/T boundary. This does not mean that what happened to vertebrates in North America at this boundary can be indiscriminately expanded to the rest of the world, but it is better to use an actual fossil record, even with its biases and gaps, in these discussions than to spin imagined patterns of extinction and survival from whole cloth.

THE NORTH AMERICAN LATE CRETACEOUS NONAVIAN DINOSAUR RECORD

Interpretations of the Late Cretaceous nonavian dinosaur record in North America remain controversial even though it is unarguably the best such record we have for this interval of geologic time. There are two main issues: (1) what was happening to nonavian dinosaurs over the last 10 million years of the Cretaceous, and (2) what was happening to nonavian dinosaurs in the last few hundred thousand years or less near and at the K/T boundary. The quality of the fossil record needed to address these two issues is quantitatively different. In the first case, one needs well-sampled, paleoecologically similar nonavian dinosaurian faunas, each spanning a few million years at most, encompassing the last 10 million years of the Cretaceous. In the second case, one needs an as-tightly-sampled-as-possible record of nonavian dinosaur faunas from the last few hundred thousand years of the Cretaceous up to and including the K/T boundary, with vertebrates on the Tertiary side to document what existed after

the boundary. The analogy I have often used in talks on the subject is a horserace. One needs to see the race up to and through the finish line as well as into the winner's circle to know who won (and who lost).

For the longer 10-million-year pattern it would be nice to have good records every half million years or so, but in reality the two nonavian dinosaurian faunas that provide the best information range from about 76–74 million years ago and about 69–66 million years ago. Before discussing the details of this I must comment on a false assumption about such a comparison. Basically, the argument is that one cannot compare two points to establish a trend. In the comparisons that ensue, it must be remembered that there are three points: 76–74 million years ago, 69–66 million years ago, and after 66 million years ago—after the K/T boundary.

At this juncture it is also worth reiterating what I mean by nonavian dinosaurs (fig. 1.1). The formal taxon Dinosauria includes two major clades, Ornithischia and Saurischia. In turn, dinosaurs' nearest extinct relatives are the pterosaurs, with the next nearest living relative being crocodilians. The best-known ornithischians (bird-hips) are stegosaurs, ankylosaurs, hadrosaurs, pachycephalosaurs, and ceratopsians. The two main branches of saurischians (reptile-hips) are sauropods and theropods. In turn, Theropoda includes not only *T. rex* and its kin but also birds. The reason for specifically using the phrase *nonavian dinosaurs* or, in some cases, *nonavian theropods,* is to designate when I am excluding birds (Aves) from a particular comparison, most notably when I am discussing extinction patterns. All these major groups of nonavian dinosaurs except stegosaurs are known from the Late Cretaceous, although in North America sauropods are rare in the Late Cretaceous and are not found north of southern Wyoming during this interval of time.

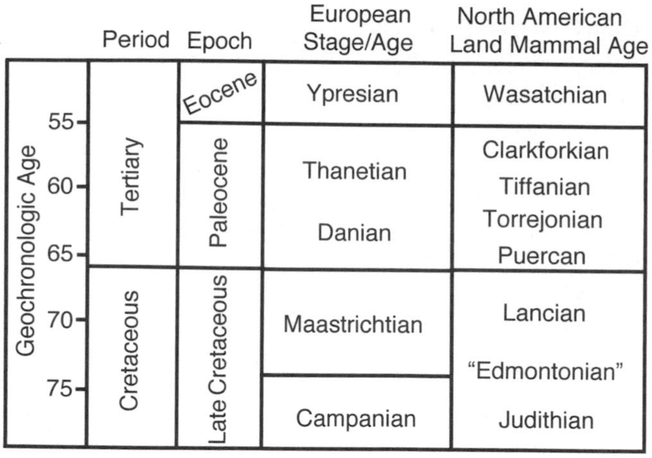

	Period	Epoch	European Stage/Age	North American Land Mammal Age
Geochronologic Age (55, 60, 65, 70, 75)	Tertiary	Eocene	Ypresian	Wasatchian
		Paleocene	Thanetian	Clarkforkian / Tiffanian / Torrejonian
			Danian	Puercan
	Cretaceous	Late Cretaceous	Maastrichtian	Lancian
				"Edmontonian"
			Campanian	Judithian

Fig. 1.2. Geologic time scale for approximately 30 million years surrounding the K/T boundary. Boundaries between North American "Land Mammal Ages" (except for the Lancian/Puercan boundary) are not yet well positioned relative to standard European stage/ages.

THE MYTH OF NO NONAVIAN DINOSAUR DECLINE DURING THE LAST 10 MILLION YEARS OF THE CRETACEOUS

Without doubt, the taxonomically richest North American nonavian dinosaur fauna comes from Dinosaur Provincial Park in southern Alberta, Canada.[9] Although the beds in the park range from about 78 to 74 million years old, most exposures are of the 76- to 74-million-year-old Dinosaur Park Formation.[10] In geologic parlance, this represents the Late Campanian portion of the Late Cretaceous. Cretaceous is at least passingly familiar to most people, but the Campanian probably is not (fig. 1.2). It is the penultimate, paleontologically defined time interval of the Cretaceous. The ultimate such interval is the Maastrichtian. Both were originally defined in Europe in the mid-nineteenth-century (Campanian, after the French hills in "Grande Champagne,"[11] and Maastrichtian, after the Dutch city Maastricht) based on fossilferous marine sections but have now been expanded worldwide. They have been modified and modernized utilizing dating techniques based on newer fossil and radiometric information. More local or regional terminology has grown up in various parts of the world. In western North America, names are applied to terrestrial, geological sections based on vertebrate fossils. For the Late Campanian, the name that has been applied is Judithian, after the Judith River Formation, first recognized in Montana.

The two major groups of dinosaurs—ornithischians and saurishcians—are very well represented in the Late Campanian (Judithian)–aged sediments of Dinosaur Park. The saurischians are limited to theropods, the group including both birds and the mostly large to extremely large ground-dwelling, bipedal, mostly carnivorous nonavian dinosaurs.[12] As noted earlier, sauropods did not make it this far north in the Late Cretaceous. Thus, the Dinosaur Park nonavian dino-

saur fauna can best be grouped as ornithischians and theropods. Of the better-preserved material, ornithischians outnumber theropods by four to one. Of these ornithischians about 50% are specimens of duck-billed hadrosaurs, 25% are horned ceratopsians, with the remaining 25% consisting mostly of the armored ankylosaurs and bone-headed pachycephalosaurs. Taxonomically, the ornithischians in the park comprise: three genera of ankylosaurs, each with one species, three species of the pachycephalosaur genus Stegoceras, six species of ceratopsians belonging to four different genera, and eight hadrosaurs arrayed in six different genera. The basal ornithischian family Hypsilophodontidae is known only from isolated teeth in the park and thus cannot be clearly assigned to a genus or species (table 1.1).

Although the theropods are much rarer, making up less than 10% of fossils recovered, they are quite diverse. According to Phil Currie, "Of the fourteen families (forty genera) of dinosaurs recognized at present in the Park, 42 percent of those families (40 percent of the genera) are theropods (excluding birds)."[13] What theropods lacked in numbers of specimens some 75 million years ago in Canada they made up in diversity. As we will see, this pattern of higher taxonomic diversity but low specimen abundance is echoed some 10 million years later, just before nonavian dinosaur extinction.

As already noted, it would be nice to have good samples of nonavian dinosaur taxa every half a million years or so, but this is unfortunately not the case. We must jump to the latest Cretaceous, specifically, the Late Maastrichtian, in order to have a comparably well-sampled nonavian dinosaur fauna. The regional name for the terrestrial vertebrates of this age, equivalent to the older Judithian, is the Lancian, named after the type Lance Formation of eastern Wyoming. The type Lance Formation may be the taxonomically most diverse Lancian-aged nonavian dinosaur fauna, with 18 species (see table 1.1). Its upper limit is essentially the K/T boundary, at about 66 million years ago or, using the newly claimed precision, 65.957 ± 0.040 (65.940).[14] An error bar of only 40,000 years for the K/T boundary is quite a claim for precision. This would be comparable to establishing the time of an event in your life in the past year to within a 10-hour window. I will use the less wieldy, but also less exact, date of 66 million years ago for the K/T boundary. The base of the Lancian is less well constrained, but may be as old as 69 million years ago.[15]

The Dinosaur Park and Lance formations are similar in several ways. First, the time represented by the Dinosaur Park Formation is about two million years, while the Lance Formation may represent an only slightly longer time, from slightly more than two million years to at most three million years. The comparable or longer length of time represented by the Lance Formation must be kept in mind as we explore nonavian dinosaur diversity between the two formations. Further, we must keep in mind that while both formations span the times indicated, they do not preserve events

Table 1.1. Generic and species counts of nonavian dinosaurs for Dinosaur Park Formation compared to Lance Formation, with additions from the Judith River, Two Medicine, and Hell Creek formations

Dinosaur Park Fm. (Judithian, Late Campanian)*		Lance Fm. (Lancian, Late Maastrichtian)[∞]		Change in # of species (with J, T, M & D)
Genus	# species (J & T)	Genus	# species (M & D)	
THEROPODA		THEROPODA		−9 (−7)
Albertosaurus (J)	(1)	Albertosaurus	1	
Bambiraptor (T)	(1)	Caenagnathus (D)	(1)	
Caenagnathus	2	Chirostenotes (M, D) Elmisaurus	(1)	
Chirostenotes	1	Dromaeosaurus	1	
Daspletosaurus	2	Nanotyrannus (M)**	(1)	
Dromaeosaurus	1	Ornithomimus	1	
Dromiceiomimus	1	Richardoestesia	1	
Elmisaurus	1	Saurornitholestes	1	
Gorgosaurus	1	Struthiomimus (D)	(1)	
Hesperonychus	1	Troodon	1	
Ornithomimus	1	Tyrannosaurus	1	
Richardoestesia	1			
Saurornitholestes	1			
Struthiomimus	1			
Therizinosaurid	1			
Troodon	1			
ANKYLOSAURIA		ANKYLOSAURIA		−1 (−2)
Edmontonia (1 is from J)	1 (+1)	Ankylosaurus	1	
Euoplocephalus	1	Edmontonia	1	
Panoplosaurus	1			
EUORNITHOPODA		EUORNITHOPODA		+1 (0)
Hypsilophodontid	1	Bugenasaura	1	
Orodromeus (T)	(1)	Thescelosaurus	1	
HADROSAURIDAE		HADROSAURIDAE		−6 (−9)
Brachylophosaurus	1	Edmontosaurus	2	
Corythosaurus	1			
Gryposaurus	2			
Hypacrosaurus (T)	(1)			
Lambeosaurus	2			
Maiasaura (T)	(1)			
Parasaurolophus (1 is from T)	1 (+1)			
Prosaurolophus	1			
PACHYCEPHALOSAURIA[††]		PACYCEPHALOSAURIA[††]		−2 (−1)
Stegoceras	3	Pachycephalosaurus	1	
		Stegoceras (M, D)	(1)	
CERATOPSIA		CERATOPSIA		−3 (−6)
Achelosaurus (T)	(1)	Diceratops (or Diceratus)	1	
Avaceratops (J)	(1)	Leptoceratops	1	
Anchiceratops[†]	1	Triceratops	2[§]	
Centrosaurus	1			
Chasmosaurus	3			
Einiosaurus (T)	(1)			
Leptoceratops	1			
Styracosaurus	1			
Total numbers of species	38 (48)		18 (23)	−20 (−25)

Sources: For Dinosaur Park Formation, Currie 2005, Ryan and Evans 2005, and Longrich and Currie 2009; for other formations, Weishampel, Dodson, and Osmólska 2004.

*Additional genera from the Judith River (J) and Two Medicine (T) formations of central Montana.

**Although here maintained as a genus, the taxonomic status of *Nanotyrannus* is equivocal, with some advocating generic status, others arguing that it is a species of *Tyrannosaurus*, and others suggesting it is a juvenile *T. rex*. Phil Currie (pers. comm. 2009) noted, "Logic says it is a young *T. rex* (why would there be only juvenile *Nanos*), but there is still some evidence to suggest it is a distinct species. I will probably go with the former interpretation if no new evidence is forthcoming." See also Currie 2003.

[†]Found south, outside of Dinosaur Park.

[††]Mark Goodwin (pers. comm. 2009) indicated that the recognition of three species of *Stegoceras* from the Dinosaur Park Formation may be obscured by the possibly juvenile status of the specimens. For the Lancian, Horner and Goodwin (2009) recognized two monotypic species, one belonging to *Stegoceras* (= *Sphaerotholus*, = *Prenocephale*) and the other belonging to *Pachycephalosaurus* (= *Stygimoloch*, = *Dracorex hogwartsia*). They wrote: "*Dracorex hogwartsia* (juvenile) and *Stygimoloch spinifer* (subadult) are reinterpreted as younger growth stages of *Pachycephalosaurus wyomingensis* (adult)" (2009: 1).

[∞]Additional genera from the Hell Creek Formation of eastern Montana (M) and western part of the Dakotas (D).

[§]The number of species of *Triceratops* is uncertain. I use the higher number of two. Further, according to Scannella (2009), *Torosaurus latus,* from the Lance Formation, represents older individuals of *Triceratops* and is thus not recognized here.

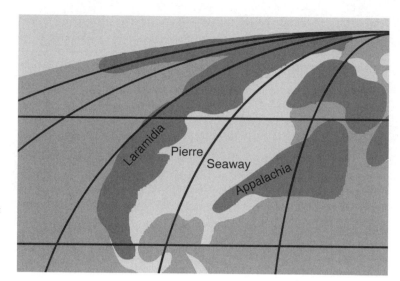

Fig. 1.3. North American paleobiogeography in the latter part of the Cretaceous Period showing the major western continent of Laramidia and the eastern continent of Appalachia separated by the shallow Pierre Seaway. In part after Smith et al. 1994.

through the entirety of the interval but rather represent time-averaged packets of time throughout the indicated intervals. Almost all terrestrially derived fossils are from such time-averaged occurrences.

Second, both formations are similar in depositional history as well, preserving coastal fluvial and floodplain deposits draining easterly into the Pierre or Western Interior Seaway. This large epicontinental sea divided North America into a western Laramidia and an eastern Appalachia for most of the 35 million years of the Late Cretaceous (fig. 1.3), transgressing (waxing) and regressing (waning) four times during these 35 million years. "Epicontinental" refers to the fact that the sea that flooded the continent was very shallow, as seas go; at its deepest it probably reached 900 m, but it was usually much shallower. The greatest widths could reach 1,000 km, with lengths up to five times that number. During the Late Cretaceous such seas inundated continents all over the globe, probably totaling an area equaling the size of modern Africa. We have nothing comparable in size today, but Hudson Bay, the North Sea, and the Persian Gulf are much smaller examples.

Turning to a comparison of the nonavian dinosaur faunas, that for the Lance Formation, while still quite rich, is no match for the older Dinosaur Park Formation nonavian dinosaur fauna. Whereas Dinosaur Park has 38 species, Lance has 18. This means a drop of about 53% species abundance in the last 10 million years of the Cretaceous in North America (see table 1.1). Some might and have questioned whether these are fair comparisons. Whereas I do not see any basis for this apprehension, other Judithian- and Lancian-aged nonavian dinosaur faunas from western North America can be added to assuage these concerns. The other major collecting areas for the Judithian age in the northern Western Interior are from the Judith River and Two Medicine formations in central Montana, while the other major areas for the Lancian age in the same region are from the Hell Creek Formation of eastern Montana and western North and South Dakota.

When these other records augment the Dinosaur Park and Lance formations, the loss of nonavian dinosaur species decreases only slightly to 52% (see table 1.1).

Not all workers have agreed that comparing the nonavian dinosaurian faunas of the Dinosaur Park Formation and Lance Formation is the most ecologically meaningful way of assessing the pattern of nonavian dinosaur diversity through the last 10 million years of the Cretaceous, going so far as calling such an assessment absurd.[16] Some have even attempted to statistically discount this very real decline, but without success.[17]

A more recent study compiled genus-richness estimates for three Mesozoic nonavian dinosaur clades. Very importantly, Barrett and colleagues conclude, "In contrast to other recent studies, we identify a marked decline in dinosaur genus richness during the closing stages of the Cretaceous Period, indicating that the clade decreased in diversity for several million years prior to the final extinction of nonavian dinosaurs at the Cretaceous-Palaeocene boundary."[18] They show that many fluctuations in nonavian dinosaur diversity through the Mesozoic found in previous studies are likely not real but are artifacts of fossil preservation, an important point that will be further explored below. These authors further found that when biases in fossil preservation are taken into account several genuine diversity signals are present, the most intriguing of these being the "diminution of ornithischian and theropod dinosaur lineages several million years prior to the K-P extinction event: both clades underwent a drastic decline."[19]

Statistical approaches to such issues are certainly justifiable when the data are poor, but this is not the case in this instance, especially after any issues of fossil preservation are eliminated. And as we see, different statistical approaches have yielded opposite results. The Late Cretaceous vertebrate fossil record is very good and by all measures is reliable in providing proportional, if not absolute, changes between Judithian and Lancian, and Lancian and K/T boundary sites. It

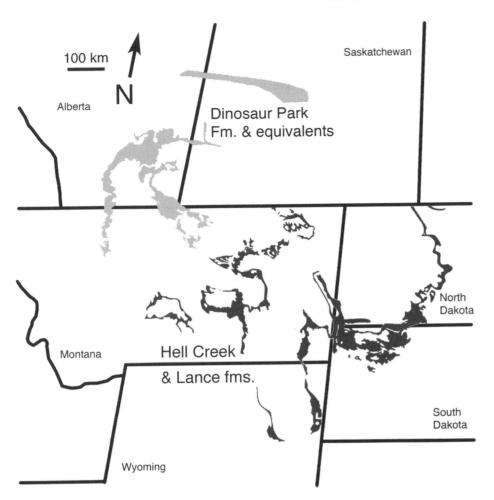

Fig. 1.4. A comparison of the mapped areas of the Campanian (Judithian) Dinosaur Park Formation and its equivalents (48,000 sq km) and the mapped areas of the Maastrichtian (Lancian) Lance and Hell Creek formations (57,000 sq km).

borders on intellectual dishonesty to lump such well-sampled localities with much poorer localities throughout the Mesozoic reign of nonavian dinosaurs.

Fastovsky and his colleagues repeat one of the common canards against comparison of the Dinosaur Park and Lance Formation nonavian dinosaur faunas: "There is no reason to suppose that the richest formations of each time interval are spatially equivalent or preserve faunas with equal fidelity."[20] By "spatially equivalent" I assume that these authors are writing that one cannot assume that these formations have similarly sized exposures. The argument here basically is that if you have more exposure, everything else being equal, you should also have more species of nonavian dinosaurs preserved. Following their reasoning, if the much greater taxonomic richness of the Dinosaur Park Formation could be attributed to a greater geologic exposure, these authors might have a case. But this is not the case.

"The bedrock exposures in the [Dinosaur Provincial] Park comprise Canada's largest area of badlands, covering approximately 75 km²."[21] The Dinosaur Park Formation constitutes a large portion of this area and, as noted above, has produced 38 species of nonavian dinosaur. If we take an even broader approach, the Dinosaur Park Formation in Alberta and Saskatchewan, plus the equivalently aged Judith River and Two Medicine formations to the south in Montana, is mapped as

covering something like 48,000 sq km[22] (fig. 1.4). This, however, adds 10 additional nonavian dinosaur taxa to the 38 already known in and close to Dinosaur Provincial Park (see table 1.1), giving a total of 48 Late Campanian (Judithian) nonavian dinosaurs from these three formations. In comparison, the Late Maastrichian (Lancian) nondinosaurian faunas from the Lance Formation of Wyoming and the Hell Creek Formation of eastern Montana and western North and South Dakota are mapped as covering at least 57,000 sq km and yet have yielded at most 23 species of nonavian dinosaurs.[23]

What this exercise shows is that Late Campanian (Judithian) nonavian dinosaur-bearing formations in western North America have less exposure than Late Maastrichian (Lancian) nonavian dinosaur-bearing formations in western North America, and yet they have yielded markedly more species. Specifically, although the Late Campanian faunas are known from 16% fewer exposures than are the Late Maastrichtian faunas, they have 45% more species. This percentage drop in diversity echoes that found by the 2009 study of Barrett and colleagues noted earlier.

The second concern that Fastovsky and his colleagues voice is that "there is no reason to suppose that the richest formations of each time interval . . . preserve faunas with equal fidelity."[24] This speaks to the issue of whether the Dinosaur Park and age-equivalent formations and the Lance and age-equivalent formations are equally rich in the number of identifiable specimens they have yielded. This cannot be addressed as easily as their question concerning spatial equivalence. There are no complete lists of the identifiable specimens that have been removed over the past 100 years and more. Many museums around the world have nonavian dinosaurs, especially from the Lance / Hell Creek formations. It is fair to say that both sets of formations have yielded countless nonavian dinosaur specimens, thus it is up to naysayers such as Fastovsky and his colleagues to demonstrate that there is a significant preservational bias between these two sets of formations. Barring this, the argument of preservational bias between the Dinosaur Park and Lance formations must be treated as a red herring.

There is another pattern of change in the last 10 million years showing that the decline in species abundance is most certainly real. It concerns the patterns of change over the last 10 million years for the far rarer theropods compared to the changes for more common hadrosaurids and ceratopsians. If the Lance nonavian dinosaur fauna were more poorly sampled than the Dinosaur Park fauna, it would show first in a depauperate nonavian theropod fauna, that is, the rarer theropods should be the first to disappear from the record if it is a matter of sampling. This is decidedly not the case. Recall that for Dinosaur Park, nonavian theropods account for only 10% of the specimens yet comprise about 41% of the species abundance. By the time of the Lance Formation, while nonavian theropods are equally rare as specimens, if anything, they may have increased taxonomically, ranging from between 39 to 48% of the nonavian dinosaur species abundance. Their numbers of species, however, have declined between 39 and 56% compared to those of theropods from the Dinosaur Park faunas (see table 1.1). This is further strong evidence indicating that comparative sampling is not an issue. The decline is real.

Yet another pattern concerns the most common nonavian dinosaurs in the Dinosaur Park and Lance / Hell Creek nonavian dinosaur faunas, the hadrosaurids and ceratopsians. Recall that for the Dinosaur Park fauna, among the ornithischians, hadrosaurs (50%) and ceratopsians (25%) account for 75% of the specimens. For Lance / Hell Creek this percentage is, if anything, higher, and yet hadrosaurids decline taxonomically by 75% and the ceratopsians declined by 43% between these two faunas. Again, there is no problem with comparative sampling. At the end of the Cretaceous in western North America hadrosaurids and ceratopsians are very common as specimens but show quite low taxonomic diversity. Hadrosaurids and ceratopsians did decline in the Western Interior of North America during the last 10 million years of the Cretaceous. If the other equivalent Judithian- and Lancian-aged nonavian dinosaur faunas provided in table 1.1 are added to this mix, the hadrosaurids and ceratopsian taxonomic declines are even more precipitous, 82% and 60%, respectively.

What these two patterns clearly show is that it was the most common hadrosaurids and, to a lesser extent, ceratopsians, rather than the rarer theropod taxa, that suffered the greatest taxonomic decline during the last 10 million years of the Cretaceous in Western North America. Such a pattern cannot be explained away by statistical machinations.

The cause or causes of these patterns of decline are harder to come by, but one possibility is that, as noted earlier, the deposition of the Lance and Hell Creek formations occurred as part of the last and most massive regression of the Pierre Seaway. The fossil vertebrate evidence available for the four transgressions and regressions in North America during the 35 million years of the Late Cretaceous suggests that during times of transgressions freshwater vertebrates suffer but land-dwelling species, including nonavian dinosaurs, fare better, even showing a taxonomic increase. The opposite occurs during regressions, and at best, land-dwelling species show taxonomic stasis, as in the Dinosaur Park Formation, or decline, as during the much greater regression that occurred when the Lance / Hell Creek vertebrate faunas lived.[25]

The changes occurring between the Dinosaur Park and Lance / Hell Creek nonavian dinosaur faunas are not unlike changes we see in extant mammalian biotas. The single Lance / Hell Creek genus of hadrosaurid, *Edmontosaurus*, is arguably but not demonstrably represented by more identifiable remains than are any of the seven hadrosaurids from Dinosaur Park. This suggests that while the number of hadrosaurid species dramatically declined in the last 10 million years of the Cretaceous, the one or two species that replaced them were very common. This is similar to the pattern of African savannah ungulates compared to ungulates of pre-European human invasion of North America. In Africa there are as many as 60 medium (above 25 kg) to large grazers and browsers (elephants, rhinos, zebra, pigs, hippos, deer, giraffes, antelope, sheep, goats, cattlelike bovids),[26] while in North America there are only 12 such mammals (peccaries, deer, elk, moose, caribou, sheep, goats, bison, pronghorns).[27] In Africa, some large migrating mammals, such as zebras and wildebeest, have prodigious numbers, but other herbivorous species are not uncommon. The migratory populations of wildebeest in the Serengeti-Mara have been measured as high as 1.3 million individuals, but single concentrations generally range from 10,000 to 20,000 animals and are usually smaller.[28] In North America, the bison was far and away the most numerous large grazing mammal when Europeans arrived some 500 years ago. In the fifteenth century 60 million bison roamed North America in massive migrations, and one estimate placed a migrating herd at over four million animals![29] By 1900 only 1,000 animals remained, but this number has now risen to more than 30,000 individuals.[30] North America had a large biomass of large ungulates, but it was dominated by a single species. Possibly a trend toward lower diversity but still with a large biomass was what transpired for large herbivorous nonavian dinosaurs in the waning 10 million years of the Cretaceous in North America.

This scenario does not explain easily why theropod (carnivore/scavenger) diversity did not decline as dramatically. One possibility is that, because the vast majority of these theropods were ostrich size or smaller and had a diversity of diets,[31] a decline in the larger hadrosaurid and ceratopsian diversity would not have affected these smaller theropods as dramatically.

A final statistically based pattern that needs to be addressed is the supposed decline of nonavian dinosaurs between the Late Campanian into the Early Maastrichtian and then the subsequent increase in the late Maastrichtian. Both supporters of an overall decline in the Late Cretaceous and those arguing against such a decline have noted this downward blip and subsequent increase. In the first camp, while arguing for a drop in taxonomic richness "between the late Campanian and the Cretaceous-Paleocene (K-P) boundary event," Barrett and his colleagues have noted "there is apparently a small rebound in diversity in the latest Maastrichtian."[32] In the second camp, Fastovsky and his colleagues acknowledge an "apparent decline in generic richness from the Campanian to the Maastrichtian, which we clearly acknowledged before . . . , and we again see the apparent increase in diversity between the early and late Maastrichtian. So we again ask ourselves, as we did when we wrote our manuscript, are these fluctuations significant or are they due to sampling?"[33]

The answer is that without looking at the actual patterns of change of the taxa through these intervals one cannot cogently argue which patterns are and are not real. As I have demonstrated with this approach, the decline in generic richness from the Campanian to the Maastrichtian is quite real, but what of the supposed dip in the Early Maastrichtian and the subsequent rise in the Late Maastrichtian? This issue cannot be answered with certainty, but once again the actual record provides a strong hint of what happened.

In an earlier, broader study, Norm Macleod and I looked at all of North America and found that the Late Campanian had 48 dinosaurian genera (including birds) that were reported from a total of 92 occurrences, while the Early Maastrichtian had only 23 genera from 32 occurrences, and the Late Maastrichtian had 32 genera from 119 occurrences (see table 1.2).[34] The first suspicion that something is wrong with the Early Maastrichtian record is that it has between 66 and 75% fewer occurrences of genera than for the time intervals before and after, respectively. The second issue is that five genera (including one bird) reported from both the Late Campanian and Late Maastrichtian are not known from the Early Maastrichtian. If these five ghost genera are included (as they should be) in the Early Maastrichtian record, there are 28 genera for this interval, only four fewer than reported for the late Maastrichtian. The dip in dinosaurian taxonomic diversity in the Early Maastrichtian is almost certainly artifactual.

THE END IS DRAWING NIGH—WITH NEITHER A WHIMPER NOR A BANG

As indicated earlier, the two main issues regarding the Late Cretaceous record of nonavian dinosaurs are (1) what was happening to nonavian dinosaurs over the last 10 million years of the Cretaceous, and (2) what was happening to nonavian dinosaurs in the last few hundred thousand years or less near and at the K/T boundary. As just discussed at length, there is little question that at least in North America, nonavian dinosaur species abundance declined by at least 50% over the last 10 million years of the Cretaceous. The second question is decidedly more difficult to tackle. To do so we must move entirely to the latest Cretaceous Hell Creek Formation and the overlying Early Paleocene Tullock Formation of eastern Montana, with a foray into North Dakota and South Dakota. This is because the records of all vertebrates, not just nonavian dinosaurs, have been more thoroughly studied in these states, the K/T boundary is better known, and at least in eastern Montana there is a reasonably good vertebrate record in the overlying Paleocene beds.

Except for one imprecise and now largely discounted study, which looked at what happened to familial level diversity of nonavian dinosaurs in a three part partitioning of the Hell Creek Formation,[35] we have had problems determining what ensued for nonavian dinosaurs and other vertebrates through the approximately three-million-year extent of the Hell Creek Formation leading up to the K/T boundary. More recently, Dean Pearson and his colleagues were able to identify more than 10,000 specimens of very small to very large vertebrates of 61 vertebrate taxa from the 100 m section of Hell Creek Formation exposed mostly in southwestern North Dakota.[36]

One major conclusion seems well justified, namely, that there is no evidence for a decline in vertebrate diversity through the vertical extent of the formation. Problems arise with their second conclusion: there is no evidence for a decline of nonavian dinosaur diversity in the three meters below the K/T boundary. In fact, there is no evidence for any vertebrate fossils in at least the last 1.77 m of their section (fig. 1.5). The single specimen that constitutes this record is identified as an indeterminate ceratopsid. Further, of the more than 10,000 specimens they collected, there are only three nonavian dinosaurs identified between 2 m and 3 m below the K/T boundary. These consist of an indeterminate ceratopsid, and indeterminate hadrosaurine, and the theropod tooth taxon *Richardoestesia* (named for my colleague at San Diego State University, Richard Estes, who died in 1990). As one moves down section, according to their figures, one encounters five nonavian dinosaur taxa just below 3 m and eight taxa just below 8 m. Does this mean that, counter to their arguments, we see a decline from eight to five to three to one and finally to zero, at 1.77 m below the K/T boundary? I think not.

The left side of figure 1.5 is based on Pearson et al.'s figure 2. On the right side I have pulled their record for nonavian dinosaurs. The numbers are the nonavian dinosaur taxa

Table 1.2. Comparison of Late Campanian, Early Maastrichtian, and Late Maastrichtian generic dinosaur diversities in the northern Western Interior

L. Campanian		E. Maastrichtian		L. Maastrichtian	
14 localities	Oc	3 localities	Oc	18 localities	Oc
Achelousaurus	1	*Albertosaurus*	2	*Alamosaurus*	7
Albertosaurus	1	*Anchiceratops*	2	*Albertosaurus*	2
Anchiceratops	1	*Arrhinoceratops*	1	*Ankylosaurus*	4
Apatornis	1	*Aublysodon*	1	***Avisaurus***	1
Aublysodon	2	*Caenagnathus*	1	*Bugenasaura*	2
Avaceratops	1	*Chirostenotes*	1	*Caenagnathus*	1
Avimimus	1	*Daspletosaurus*	1	*Chirostenotes*	2
Avisaurus	1	*Dromaeosaurus*	2	*Diceratops*	1
Bambiraptor	1	*Edmontonia*	3	*Dromaeosaurus*	7
Baptornis	2	*Edmontosaurus*	3	*Edmontonia*	6
Brachyceratops	1	*Euoplocephalus*	1	*Edmontosaurus*	9
Brachylophosaurus	3	*Hypacrosaurus*	1	***Leptoceratops***	3
Centrosaurus	2	*Maiasaura*	1	*Montanoceratops*	1
Chasmosaurus	1	*Montanoceratops*	2	*Nanotyrannus*	1
Chirostenotes	2	*Ornithomimus*	1	*Ornithomimus*	5
Coniornis	1	*Pachyrhinosaurus*	2	***Pachycephalosaurus***	4
Corythosaurus	1	*Panoplosaurus*	1	*Pachyrhinosaurus*	1
Daspletosaurus	3	*Parksosaurus*	1	*Palintropus*	1
Dromaeosaurus	4	*Richardoestesia*	1	*Parksosaurus*	1
Dryptosaurus	1	*Saurolophus*	1	***Pentaceratops***	1
Edmontonia	4	*Saurornitholestes*	1	*Potamornis*	1
Einiosaurus	1	*Stegoceras*	1	*Richardoestesia*	3
Euoplocephalus	3	*Struthiomimus*	1	*Saurornitholestes*	6
Gorgosaurus	2	**23 genera**	**32**	*Sphaerotholus*	1
Gravitholus	1			*Stegoceras*	2
Gryposaurus	2			*Struthiomimus*	1
Hadrosaurus	1			*Stygimoloch*	4
Hesperornis	2			*Thescelosaurus*	6
Hypacrosaurus	2			*Torosaurus*	8
Lambeosaurus	1			*Triceratops*	9
Leptoceratops	2			***Troodon***	6
Maiasaura	1			*Tyrannosaurus*	12
Monoclonius	2			**32 genera**	**119**
Montanoceratops	1				
Ornatotholus	1				
Ornithomimus	3				
Orodromeus	3				
Pachycephalosaurus	2				
Panoplosaurus	1				
Parasaurolophus	2				
Pentaceratops	1				
Prosaurolophus	2				
Richardoestesia	4				
Saurornitholestes	6				
Stegoceras	4				
Struthiomimus	1				
Styracosaurus	1				
Troodon	5				
48 genera	**92**				

Source: Modified after Archibald and MacLeod 2007.

Note: The Early Maastrichtian has many fewer genera, occurrences (columns labeled "Oc"), and localities. Further, five genera (in **boldface**) are found in the Late Campanian and Late Maastrichtian, but not in the Early Maastrichtian.

either collected at that level or implied by the lines between bounding levels. I have also shown this visually in a spindle- or minaret-shaped diagram. If this record is to be believed, there is relative taxonomic stability until about the four-meter mark, when numbers start to drop, reaching zero before the K/T boundary. Thus, the record would seem to indicate a gradual disappearance of nonavian dinosaurs in these authors' sections. Pearson and his colleagues argue that this is artifactual and that in reality this nonavian dinosaur record is commensurate with a catastrophic extinction. Theirs is an ar-

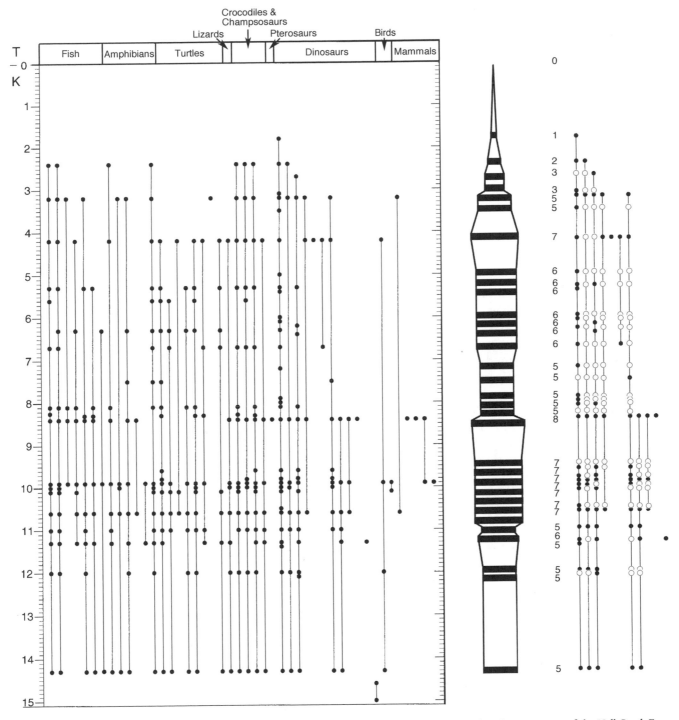

Fig. 1.5. *Left:* Isolated occurrences of taxa (dots) are identified within clades connected by vertical lines within the upper 15 m of the Hell Creek Formation, exposed mostly in southwestern North Dakota. Modified after Pearson et al. 2001. *Right:* A spindle diagram and a repeat of occurrences of dinosaurs showing numbers of dinosaurs at the various levels, interpolating the presence of a taxon if it was recovered above and below that level (open circles). Note the decline from seven to zero taxa in the last 4 m and the total lack of any vertebrates in the uppermost 1.77 m.

gument mostly derived from what is called the Signor-Lipps effect, which states that because the fossil record is never complete we should not expect to have found the first or the last fossil records of any particular species or higher group of organisms. Invoking the Signor-Lipps effect means, however, that one can claim that catastrophic, gradual, or some other pattern of extinction is commensurate with the fossil record.

I agree with Pearson and his colleagues that "it is possible that this gap is a function of depositional conditions associated with the formational transition rather than a reflection of true terminal Cretaceous faunal diversity."[37]

For those claiming that the fossil record demonstrates a catastrophic extinction of nonavian dinosaurs, there is a cautionary tale from recent studies of late Quaternary extinctions of large mammals by James Haile and his colleagues.[38] As with the K/T extinctions, some have claimed that the late Quaternary megafaunal (predominately larger mammals) extinctions were catastrophic. The claims have been made that because of human hunting, most larger mammal species, such as mammoths, disappeared in a very short time or that an extraterrestrial impact resulted in near-instantaneous extinctions. Based on the fossil record, the latest occurrences of woolly mammoths and horses in northwestern North America date from about 15,000 to 13,000 years before the present. Based on the presence of ancient DNA from perennially frozen and well-dated sediments, Haile and his colleagues found that woolly mammoths and horses persisted in the interior of Alaska until at least 10,500 years before the present, which is several thousand years later than indicated from the fossil record. Their findings show that mammoths and horses overlapped with humans for several millennia in the Americas, challenging the hypotheses that these megafaunal extinctions occurred within centuries of human arrival or were caused by an extraterrestrial impact in the Late Pleistocene. Thus, the claims that the last fossils of dinosaurs mark a catastrophic extinction must be viewed skeptically, given that reduced populations may have survived long after the record of the last fossils.

There are other problems with the somewhat simplistic interpretations of Pearson and his colleagues. If the single indeterminate ceratopsid noted above is discounted, the 1.77 m becomes a 2.37 m fossil gap in their data set. This gap has also sometimes been referred to as the "3 m" or "10 ft" gap. This notoriously unfossilferous gap was first thought to represent the level at which nonavian dinosaur extinction occurred. It is, however, below the K/T boundary, as based on the much more easily and better-sampled plant pollen record and the iridium-enriched level thought to mark the Chicxulub impact. Even before the advent of the impact theory of extinction, it was realized that this is a generally barren zone for any vertebrate fossils.

Not surprisingly, this has created considerable problems in gauging what was happening not just to nonavian dinosaurs but also to all vertebrates during this nearly barren interval. Some, such as Pearson and his colleagues, have tried to argue this gap away, basically saying that because no decline in vertebrate diversity can be detected, no such decline, especially for nonavian dinosaurs, can be detected in the three meters below the K/T boundary.[39] Given that this 3 m (or 2.37 m) gap is by far the largest gap in the upper 12 m of the Hell Creek Formation, such a claim is not warranted. There is no evidence supporting a catastrophic (bang) or gradual (whimper) extinction in these last two to three meters of the Hell Creek Formation.

This unfossilferous gap is vexing. Why is there such a gap at all? The views vary, and frankly, all explanations seem to have some merit, but I am unable to choose one over another and have some reservations about all the explanations. One

suggestion is that a lack of channel deposits within this three-meter interval may explain the absence of fossil localities.[40] The vast majority of both small and large fossil vertebrates in the Hell Creek Formation are associated with fluvial or riverlain deposits. These can be the channels themselves, sand bars that breached the channels during flood events, or more rarely, floodplain deposits on the margins of the streams. Especially compared to the overlying Paleocene Tullock (or Fort Union, as it is referred to in the Dakotas) Formation, the Hell Creek Formation tends to preserve smaller streams or at least muddier streams in its upper reaches, although very large channels certainly are not present. The upper reaches of the Hell Creek Formation for the most part lack larger channel deposits, although most of the sediment still appears to be fluvially related, and thus the rarity of channel deposits to explain the absence of fossil localities in the last three meters seems to have merit.

Another suggestion is that as a result of the Chicxulub impact at the K/T boundary the resulting global acid rain fallout likely caused the destruction of microfossils immediately underlying the impact layer.[41] This is an appealing idea, but, as will be discussed under extinction mechanisms, the survival of most aquatic vertebrate species belies the idea of a very low pH acid rain. The possibility of the dissolution of most teeth and bones in the upper three meters of the Hell Creek Formation may yet prove to be true, but not owing to acid rain. Rather, the carbonaceous and lignitic coals that most often overlie the top of the Hell Creek Formation could have caused the leaching near its top. I tend to favor this admittedly more prosaic but, I think, likelier cause. It can only be tested by finding and thoroughly exploring other terrestrial K/T sections to see if this gap is a regionally driven phenomenon or is global, as some authors argue.

Another approach to examining what was occurring leading up to and across the K/T boundary has used large samples of vertebrate microfossils amassed over almost 40 years by crews from the Museum of Paleontology at the University of California, Berkeley, working in the Late Maastrichtian (Lancian) Hell Creek Formation and Early Paleocene Tullock Member of northeastern Montana. One of the major players in this ongoing study is Greg Wilson, who began this project while a doctoral student of Bill Clemens at UC Berkeley. I agree with Wilson's assessment that, while the evidence of an impact of an asteroid or other such body at the K/T is substantial, "the causes, timing, and extent of the associated mass extinction remain unclear."[42]

To begin to address these shortcomings, Wilson and his co-workers placed vertebrate localities into a temporal framework utilizing stratigraphic position. They noted changes in the better-sampled amphibian, turtle, and mammal paleocommunities approaching the K/T boundary. They compared these data with those for published paleofloral and paleoclimatic data. Wilson has coordinated his work with Jack Horner's Hell Creek Project from 1999 to 2008 to exam-

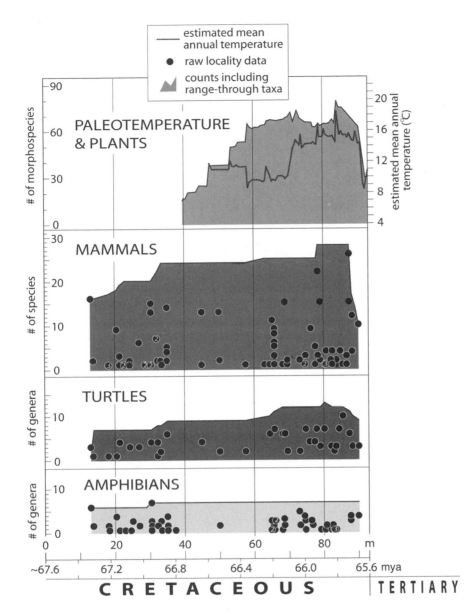

Fig. 1.6. Comparisons of paleotemperatures with diversities of turtles, mammals, amphibians, and plants during the last two million years of the Cretaceous. Note the warming trend and the correlation of a drop in taxonomic richness of all but the amphibians during the rapid drop in paleotemperature in the last 100,000 years of the Cretaceous. Diagram © Greg Wilson.

ine vertebrate diversity throughout the whole of the thickness of the Hell Creek Formation in eastern Montana. While Horner's work concentrated on nonavian dinosaurs, Wilson and others looked at nondinosaurian vertebrates.[43]

Echoing the findings of Pearson and colleagues, Wilson and colleagues found "no evidence of long-term changes in paleocommunities leading up to the K-T boundary, although some changes in relative abundances, taxonomic composition, and body sizes likely reflect normal responses to background levels of climate change." But unlike Pearson and colleagues, Wilson and his colleagues found that "dramatic changes in mammal and turtle paleocommunities occur within the last 100 thousand years of the Cretaceous." They go on to say while their results are "consistent with a sudden extinction from an extra-terrestrial impact; non-linear response to long-term causes or multiple short-term causes

cannot be rejected. Future work will improve the temporal resolution of the chrono-stratigraphic framework, develop a local climate signal, improve the density of fossil sampling, incorporate other fossil taxa (plants, mollusks), and draw comparisons with other study areas."[44]

This work shows that during the last two million years of the Cretaceous, maximum diversities of turtles, mammals, and plants correlate with the maximum latest Cretaceous warming trend, while the drop in taxonomic richness of these three groups correlates to a rapid drop in paleotemperature in the last 100,000 years of the Cretaceous of the same region (fig. 1.6).[45] If verified by further studies, these results clearly point to factors driving climatic change well before any extraterrestrial impact. These will be explored in Chapter 5. Clearly, extinction theories, like life, are not as simple as they might first appear.

2

In the Shadow of Nonavian Dinosaurs

WHEN I EXPLAIN TO PEOPLE that I study mammals that lived with the dinosaurs I usually receive one of two reactions, accompanied by quizzical looks. Either they had no idea that mammals lived at the time of dinosaurs or they are visualizing monkeys, horses, and humans cavorting with dinosaurs. I quickly dispel this notion, noting that apart from B movies and creationist museums, the mammals that existed with dinosaurs did not belong to groups of modern mammals. With a very few exceptions, none of these Mesozoic mammals was very large—more shrew-to-rat size, though on one rare occasion a small-dog-sized Cretaceous form named *Repenomamus* was found with the bones of a small dinosaur in its gut region.[1] On the uncommon occasion when the press writes about these ancient mammals they are referred to as shrew- or ratlike in appearance. Superficially, this is not a bad analogy; these creatures had fur, most had four rather generalized limbs and feet, and if glimpsed, they could have been seen scampering among the vegetation or scurrying down a hole. But in detail these creatures were quite different. Almost all of them had nothing to do with the mammals that behave in this same manner today, nor did they bear a close resemblance to, or relationship with, monkeys, horses, or humans.

Let us start from the beginning and define a few useful terms. As noted in the last chapter, Dinosauria is a monophyletic group or clade of vertebrates that includes what we usually think of as dinosaurs (*T. rex, Triceratops, Stegosaurus,* etc.), but it also includes birds, or more formally, Aves. When we say that a group is *monophyletic* we mean that it includes the most recent common ancestor of all organisms in that group and all descendants of that common ancestor. A monophyletic group (more correctly, a monophyletic taxon) is the same thing as a *clade,* which is a more precise way of talking about a lineage of some group of animals or plants. Aves is a special kind of clade, or monophyletic group, called a *crown clade.* This is a very appropriately descriptive term because Aves "crowns" Dinosauria as the only living clade of Dinosauria. In the same vein we can speak of all extinct nonavian dinosaurs as being stem groups relative to Aves, in much the same way that there are branches or stems lower on plants that grew before the branches at the crown of that plant.

Aves, along with crocodilians, lizards, snakes, turtles, and a few smaller groups, constitutes the living members of a group we call *Reptilia,* which also contains a host of extinct species, including the nonavian dinosaurs. For mammals there is no comparably familiar term that includes both living mammals and their more ancient, extinct relatives. The comparable, but less familiar term for living mammals plus their more ancient, extinct relatives is *Synapsida.* More will be said about synapsids later in the chapter.

Also for mammals there is not a familiar term comparable to *Dinosauria*. The comparable, but unfamiliar formal term we use to include the crown group Mammalia and all its stem groups is *Mammaliaformes*.[2] Unless I am specifically and formally referring to Mammaliaformes or Mammalia, I will use the vernacular—*mammal*.

Dinosaurs are first seen in the fossil record about 228 million years ago, in the Late Triassic, and mammaliaforms are first seen about 220 million years ago, also in the Late Triassic. Within the context of the billions of years of life on planet Earth and the vagaries of the fossil record, the time of appearance of these two clades is essentially the same. If one measures evolutionary success by looking at who are the largest and most numerous species at any given time in the geological past, both nonavian dinosaurs and mammaliaforms had their heyday.

Dinosaurs included the largest-ever land animals and were moderately numerous from 228 million years ago until all but the bird clade perished at the end of the Cretaceous, 66 million years ago. From 66 million years onward, mammaliaforms in the guise of mammals quickly came to dominate in numbers of species as well as including the largest land animals in the Cenozoic Era and by far the largest animals that ever lived—cetaceans. Dinosaurs, in the guise of birds, did survive the extinction that befell most other nonavian dinosaurs 66 million years ago. At between 9,000 and 10,000 species, birds outnumber species of mammals almost two to one.[3] In terms of ecological diversity, however, birds do not compare. In size, birds range from hummingbirds to ostriches, while mammals range from shrews to whales—a difference in scale of many magnitudes. In reproduction, all birds are oviparous or lay eggs, while mammals exhibit all three major kinds of reproduction known to vertebrates: oviparity (monotremes or egg-laying); ovoviviparity, in which the egg is retained by the female, who gives birth to a very immature young (marsupials); and euviviparity, in which the developing embryo is retained and receives all sustenance from the female (placentals).

In form, birds certainly vary widely, but all have two hind limbs, for grasping or propulsion on land or in the water, and two forelimbs, which are usually for flight but are sometimes modified for swimming or can be greatly reduced; they are never for grasping or running on land. In contrast, early mammals retained a more generalized pattern of limb development that did not lock them into the avian forelimb–hind limb dichotomy. As a result, although there are only about 5,400 living species of mammals,[4] there is a tremendous amount of diversity in limb form and function.

From the basic five digits of both fore and hind limbs have arisen paddles, wings for flying, membranes for gliding, tools for digging tools, hoofs for fleet-footedness, deadly claws for capture of prey, and fingers to play a Mozart piano sonata. Accompanying this is an equally staggering variation in forms of the mammalian tools for food acquisition—teeth. There are teeth for nipping plants, slicing through meat, scrapping sap from trees, grinding down tough plant materials, or even pushing over trees in the case of elephant tusks. Mammals also repeatedly lost all their teeth when going after termites and ants, or evolved a new substance called baleen, for filtering microscopic organisms for the ocean. I have more to say on the evolution of teeth later.

A HISTORY OF MESOZOIC MAMMALIAN HISTORY

Ask most six or seven year olds and they can rattle off many kinds of nonavian dinosaurs and pronounce the names correctly in the bargain—but not so with extinct mammals. Many of us who work on groups of fossil organisms other than nonavian dinosaurs have rolled our eyes and groaned when we read yet another press story about some scrappy pieces assigned to some new nonavian dinosaur species. A little jealous, yes, but if extinct mammals had the cachet of nonavian dinosaurs, there would probably be no reason to write this book. If extinct mammals from the Age of Mammals, dating from 66 million years ago, are less well known than nonavian dinosaurs, then mammals and mammaliaforms from the Age of Dinosaurs are downright obscure.

Why, then, am I interested in them? And why have I spent much of my almost 40-year scientific career studying them? For many scientists the reason for studying any particular subject comes down to their fascination with the subject matter. For those who do medical research, there is often the desire to help others. It is no surprise that such research takes the lion's share of funds for scientific research. Even with the public fascination with dinosaurs, the study of past or even present life receives paltry sums compared to many other sciences, notably health care. Thus, I certainly don't study the most ancient mammals for money or glory. The simplest and most succinct answer is that I am enthralled by the history of life on this planet in general and the history of mammals in particular. More to the point, I teach an advanced course on mammals that encompasses most aspects of their biology, yet when we look at the great diversity of mammals that we usually encapsulate as mammalian orders, we have no good idea whence they came.

With no exceptions, the orders of mammals that we recognize as bats, whales, carnivores, hoofed herbivores, elephants, primates, rodents, and so forth, are known in the fossil record only after the extinction of stem dinosaurs 66 million years ago. This enigma has so intrigued me that I have searched in rather remote areas in both in North America and Asia for the harbingers of the major clades of mammals. For some of us, fieldwork is fun, and the thrill of finding something never before seen, even if it is an itty-bitty mammal jaw, is a natural high that cannot be surpassed.

The first-ever discovery of these obscure little Mesozoic mammaliaforms did not occur in some faraway place like Mongolia but in 1764 in southern England, at Stonesfield, which also produced some of the earliest known dinosaurs, including the first described, *Megalosaurus*. The site is now known to be Middle Jurassic in age, roughly 170 million years old. The bits of jaw and teeth were not recognized as being from a mammal (or what we now call a mammaliaform) until 1818, when the great French anatomist and father of vertebrate paleontology, Georges Cuvier, examined the specimen at Oxford University.[5] As Tim Rowe aptly described the event, "Renowned for his ability to judge the nature and affinities of an extinct animal from a part or even a single fragment of a skeleton, Cuvier pronounced the Stonesfield specimens to be mammalian. He lived up to his reputation, and his identification was the first of many violations of what had been considered a very general rule—that mammals did not live during the Age of Reptiles."[6]

The very existence of Mesozoic mammals (now mammaliaforms) was generally not accepted until the great British anatomist Richard Owen, first friend and later nemesis of Charles Darwin, published a paper on them in 1841.[7] In the same year he spoke about, but did not name, Dinosauria in a lecture at a meeting of the British Association for the Advancement of Science.[8] He did not formally use *Dinosauria* until it was published in 1842.[9] So not only did dinosaurs and mammaliaforms show up in the fossil record nearly at the same time, more than 220 million years ago, but with some irony, Dinosauria was named and Mesozoic mammals were first accepted within a year of each other, although this acceptance was by no means universal.

The idea that there were Mesozoic mammals and a newly recognized group of reptiles called dinosaurs was used both for and against the emerging notions of evolution that were swirling around learned circles in England. Owen tried to present dinosaurs as more mammal-like, which the four-legged reconstructions of the earliest dinosaurs still show in Sydenham Park, London, to this day. The general view of the time was that if evolution had occurred, one should see the progression of more primitive to more advanced forms. Owen argued that there was no such progression in the fossil record.[10] He offered as proof his reconstructions of rather advanced mammal-like dinosaurs in the Mesozoic, which, according to his views, are today represented only by degenerated reptiles.

In this same vein of "no progression," Owen recognized modern groups of mammals in the Mesozoic. He identified *Phascolotherium*, one of the Middle Jurassic taxa from Stonesfield, as a marsupial. He did this in part because he thought that the back part of the lower jawbone (dentary) was turned inward, as in living marsupials. We know he was wrong in this assessment, although today *Phascolotherium* is now argued to be an early member of the crown group Mammalia.[11] As I discuss in Chapter 3, although it is widely accepted that the crown clade Mammalia was certainly present by the Cretaceous, if not the Jurassic, recognizing fossil members of

living clades of mammals in the Mesozoic remains a tricky proposition.

Others in nineteenth-century England held ideas of Mesozoic mammals that to us appear even more outlandish than Owen's. The influential nineteenth-century British lawyer-turned-geologist Charles Lyell had what we would deem a bizarre idea about Mesozoic mammals. Lyell championed the idea of uniformitarianism,[12] which had been argued by an earlier geologist, James Hutton, who is credited with the poetic and descriptive phrase "no vestige of a beginning, no prospect of an end" to explain the history of the Earth. Essentially for Hutton and then Lyell, not only had the Earth existed for all time but so also had all the major groups of organisms, including mammals. Thus, Lyell certainly accepted the idea that there were Mesozoic mammals—and that there should be Paleozoic mammals as well! Horses in the Silurian, anyone?[13]

Charles Darwin had been sold on Lyell's uniformitarian ideas and the idea of observing and testing nature when he took the first of a soon-to-be three-volume set of geology books by Lyell on his travels on the H.M.S. *Beagle* in 1831. Lyell's ideas greatly influenced not only Darwin's geological studies on this voyage but also his general outlook on the natural world. As Darwin began to share his ideas on evolution with Lyell in the mid 1800s, Lyell was reluctant to accept evolution or Darwin's mechanism—natural selection. Of course, Lyell's Paleozoic mammals were never found, but by the last third of the nineteenth century Mesozoic species were clearly known but were not like those from the Cenozoic; they were considered to be more primitive. They, like their contemporaries, the dinosaurs, first appear in the early part of the Mesozoic (the Triassic), both evolving over time. In the realm of biology at least, Lyell had to abandon his Huttonian notion of "no vestige of a beginning, no prospect of an end." There was a succession or supposed progression of life, which was now know to be evolution.

One of the greatest champions of the evolutionary cause, Thomas Henry Huxley, also held the earlier Lyellian idea of Paleozoic mammals, probably more accurately called proto-mammals, into the 1880s, well after Darwin's evolutionary ideas had been published. Their ideas did differ in that Lyell, in his earlier work, envisioned ancient lineages that showed no progressive change, while Huxley held that changes had occurred. In Huxley's view, one line had lead from fish to amphibians to reptiles and finally to birds, while the other went from amphibians to mammals, bypassing reptiles.

So-called mammal-like reptiles were being discovered, most notably from Triassic rocks in southern Africa, and were being described by Richard Owen. These were the perfect bridge between amphibians and mammals, yet Huxley, for somewhat unclear reasons, rejected Owen's southern African forms as the intermediates between amphibians and mammals.[14] These "mammal-like reptiles" are correctly referred to as Synapsida, as these forms have nothing to do with reptilian evolution. The name *Synapsida* comes from the arch of bone below the single opening on the side of the braincase (fig. 2.1). The remnants of this arch persist today in most mammals as the zygomatic arch and form the bone below and to the side of our eyes—basically our "cheek" bone. Thus, Synapsida is the larger clade, which includes Mammaliaformes and in turn includes Mammalia.

The irony is that Huxley's view of the origin of the clade leading to mammals was essentially correct. The lineage leading to mammals did not go through a stage that we now recognize as reptiles; rather, their ancestors and those of fossil and living reptiles (including birds) split from one another some 300 million years ago. This occurred shortly after amniotes appeared in the fossil record. Amniotes had the capabil-

ity of laying their eggs on land without the need for standing water or a moist hideaway. They could do this because they had, as do all amniotes today, three extra-embryonic membranes—the amnion, the allantois, and the chorion. Respectively, these membranes keep the embryo wet and cushioned, exchange gases and handle liquid waste, and aid in gas exchange and can, in mammals, aid in the transfer of nutrients from mother to embryo.

By the last third of the nineteenth century most scientists as well as the educated populace accepted evolution (although Darwin's mechanism of natural selection was going out of favor), and the presence of Mesozoic mammals had been well accepted for almost 30 years. The first monographic treatment specifically dealing with Mesozoic mammals appeared in 1871, written by the same person, Richard

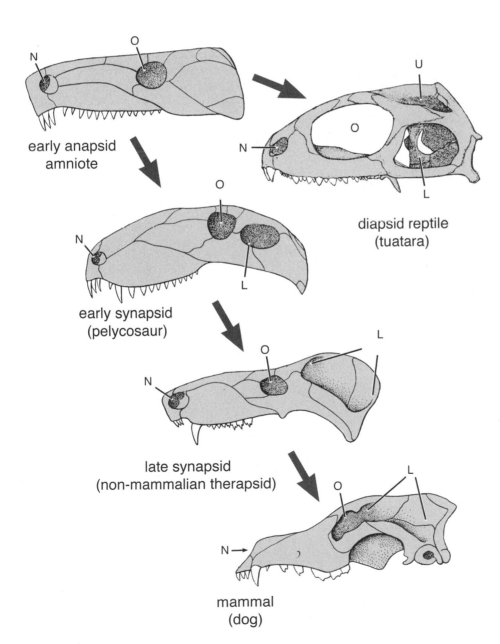

early anapsid amniote

diapsid reptile (tuatara)

early synapsid (pelycosaur)

late synapsid (non-mammalian therapsid)

mammal (dog)

Fig. 2.1. Skulls of synapsids. *Top left to bottom right:* An early amniote, a pelycosaur, a non-mammalian therapsid, and a mammal. *Top right:* A diapsid skull, found in the majority of reptiles (including birds), which has two openings in the skull roof compared to one in synapsids. Note the enlargement of the opening on the side of the skull, going from top to bottom. Abbreviations: O = orbit; N = narial opening (not seen from the side in mammals); U = upper temporal opening; L = lower, or only, temporal opening in synapsids. Not to scale. Modified after Romer and Parsons 1977.

Owen, who had firmly established the existence of Mesozoic mammals some 30 years earlier. He recognized 34 of what today we would generally equate with species, which were arrayed in 16 genera. Of the 16 genera, one is now known to be a non-mammaliaform synapsid that is a stem relative just outside of Mammaliaformes, 11 are still valid genera, and only four have been shown to not be valid genera.[15] This is quite a record for genera, some of which were named well over 150 years ago.

Where Owen and others at the time were wrong is in how these Mesozoic mammals related to living mammals. For two of the genera, Owen was not sure, but for the other 14 he thought he had marsupials. Recall that marsupials are the large group of living mammals, such as opossums, koalas, and kangaroos, and their most recent common ancestor. The perception in Owen's time, and one that unfortunately continues today, at least for the general public, is that marsupials are more primitive than and inferior to placental mammals. I recall viewing a kangaroo exhibit at our world-class San Diego Wild Animal Park some years ago where I read a sign stating that marsupials (including kangaroos) were more primitive and appeared before placentals. Marsupials did not appear before placentals and marsupials did not evolve into placentals; they are each other's nearest relatives, or sister clades, which appeared at the same time—but more on this later.

One of the major reasons that Owen regarded most of these Mesozoic mammals as marsupials is because the back part of the lower jaw, which in all mammals is composed of a single bone on each side of jaw, is markedly bent inward. This is known as an inflected dentary angle. Many placentals also have a clearly demarcated dentary angle, but it is seldom or only slightly inflected. If you feel the lower back corner of your jaw, you are touching the dentary angle, which in humans is quite reduced. As it turns out, it is the inflected dentary angle as seen in marsupials that is the more evolved (derived) condition. A superficial part of the medial pterygoid muscle has been enlarged and plays an important role in chewing in marsupials.[16] The inflected angle provides an area of insertion for this muscle (fig. 2.2).

Some other contemporaries of Owen thought some of these Mesozoic mammals were best aligned with extant placentals, but they were equally wrong. As we will see, some of these Mesozoic species are very early branching stems on the mammaliaform tree, while others were harbingers of what was to come in true, or crown, mammals.

This realization began to take form later in the nineteenth

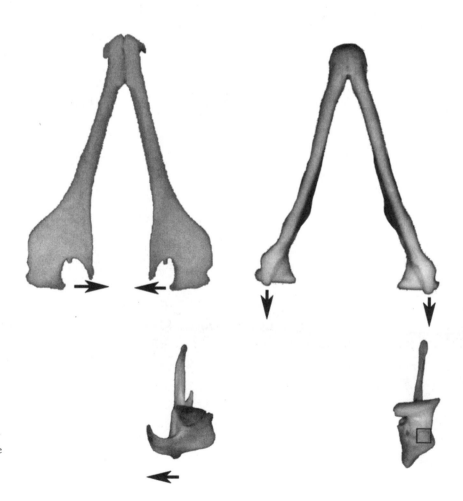

Fig. 2.2. Marsupial mandible *(left)* and placental mandible *(right)*. Ventral views *(above)* and posterior views *(below)* of the right part of the mandible. The inflected angles of the marsupial are indicated by the direction of the arrow. Posterior facing angles of the placental are indicated by the direction of the arrow (and the square). Photographs from Sánchez-Villagra and Smith 1997.

Fig. 2.3. The ceratopsians *Triceratops (above)* and *Protoceratops (below)*. The former was about 8 m and the latter almost 2 m.

century and continued into the early twentieth century, especially after the discovery of many more species of Mesozoic mammals in the United States and in Mongolia. While work during this time brought many new specimens, there was not a true synthesis of Mesozoic mammals until the publication of two important monographs in 1928 and 1929, by the most important paleontologist of the twentieth century, George Gaylord Simpson. Not only did Simpson contribute greatly to our understanding of mammalian paleontology and evolution, but he was also a major contributor to the rise of our ideas on evolutionary theory, which we call the Modern Synthesis because it combined Mendel's work on genetics with Darwin's idea of natural selection.

Simpson's 1929 volume was the published version of his dissertation and covered Jurassic and Cretaceous mammals that had been collected during the latter part of the nineteenth and the beginning of the twentieth century in the Western Interior of North America. The 1928 volume was based on postdoctoral research that Simpson conducted on European Jurassic and Cretaceous mammals that had been recovered beginning in the earlier part of the nineteenth (including those monographed by Owen in 1871) and into the early twentieth century.

The next milestone for Mesozoic mammals occurred by accident. Between 1922 and 1928, the charismatic Roy Chapman Andrews (one possible model for Indiana Jones)[17] of the American Museum of Natural History lead five well-financed but nevertheless rough expeditions into the Gobi Desert of Mongolia. They were seeking fossils documenting human evolution, but instead they found some of the most exquisitely preserved dinosaurs that had ever been unearthed. Most notable among these were those of *Protoceratops*, a very small precursor of the massive *Triceratops*, which tramped the low coastal plains of North America in the waning days of the Cretaceous (fig. 2.3). Enough was known of this dinosaur to reconstruct the ontogeny, or life history, of these creatures. It has also been surmised that *Protoceratops* might be the source of the ancient myth of the griffin because of its beak being misinterpreted as that of a bird and its frill that of a griffin's stylized wings.[18] The Late Cretaceous beds in Mongolia that produced *Protoceratops*, along with many other iconographic dinosaurs, are estimated to be about 75 million years old, some 10 million years before the demise of the nonavian dinosaurs.

While there was excitement among paleontologists when very small remains of mammals were found nestled among the dinosaurs, this did not fire the public's imagination as had the dinosaurs and the first recognized nonavian dinosaur eggs. Nevertheless, one of these mammals, *Zalambdalestes*,[19] named in 1926 by G. G. Simpson and the superb anatomist W. K. Gregory, is still recognized and is one of the more completely known Cretaceous mammals. The name means "thief with the zalambdodont tooth," and although somewhat euphonious, it unfortunately refers to a condition of the teeth known as zalambdodonty, which this mammal does not possess. The original specimens of *Zalambdalestes* were rather worn, yielding the understandable but mistaken attribution of zalambdodonty. This mistake was realized when better-preserved specimens were found—but more on *Zalambdalestes,* and its purported relationship to rabbits and rodents, later.

This zalambdodont condition evolves when one cusp on each upper molar is lost to form a large V shape of the upper molar—handy for slicing insects. The name is from Greek and roughly means "very ridged tooth." Far more common is a W-shaped tooth (dilambdodonty, or "two-ridged teeth") created by retaining the two upper molar cusps. Further, the Greek letter lambda (Λ) sometimes resembles each of the winged upper molar cusps, the paracone and metacone. We humans are primitive in retaining these two cusps in our upper molars but advanced in not retaining ridges for slicing the carapace of a beetle; instead, our cusps are low and rounded—better for chomping cheeseburgers and gnashing chips.

The next group of expeditions to the Gobi did not occur until after World War II. Soon thereafter Soviet-Mongolian, then Polish-Mongolian, and finally, continuing to this very day, Americans returned, again from the American Museum of Natural History, to follow in Chapman's footsteps. Today, groups from a number of different countries crisscross the Gobi Desert. The Poles, and increasingly the Americans, have published extensively on what can only be called some of the most-complete Cretaceous mammals in the world. I comment further on some of these fossils below and return to these finds in later chapters.

Almost in tandem with the Late Cretaceous mammalian work in Mongolia in the 1950s onward, there was renewed interest in the Late Cretaceous mammals of western North America. Here the fossils were nowhere nearly as complete as the Mongolian taxa, but there were more of them, and there were more localities showing greater detail about these Cretaceous mammalian faunas in both time and space. The reports of these species appeared in a series of shorter papers, but more notably in longer monographic treatments starting in the early 1960s.[20] Some of these mammalian faunas came from the same beds—the Hell Creek Formation—from which the type specimen of *Tyrannosaurus rex* was recovered at the very beginning of the twentieth century. Again, you can guess who received more press.

By the 1970s Cretaceous mammals turning up in what the Soviets called Middle Asia, which essentially comprises the present countries of Kyrgyzstan, Tajikistan, Turkmenistan, Uzbekistan, and southern Kazakhstan. This earlier work was almost exclusively that of the Lev Nessov and his students.[21] Most notable were the discoveries in the now independent country of Uzbekistan. These Cretaceous mammalian faunas were older than most sites in both Mongolia and North America, and they ranked in between the two in the quality of preservation.

For the Cretaceous, except for the newer work in Uzbekistan, much of the current knowledge of Mesozoic mammals was collated in an edited volume in 1979, which was for the time state of the art.[22] But things did not stand still for long.

CRETACEOUS EUTHERIAN MAMMAL DISCOVERIES POST 1980

In 1980 the Alvarez group published its hypothesis that the impact of an asteroid caused the catastrophic extinction of nonavian dinosaurs. Although nonavian dinosaurs took center stage, as always, this time it was soon realized that other species of plants and animals lived at the same time as these last nonavian dinosaurs. Although far too many papers that followed touting this or that physical corollary event of an asteroid impact by and large ignored the fossil record or only gave lip service to nonavian dinosaurs, many thought-provoking papers began to appear examining the entire fossil record through this interval. Some supported the impact as the sole cause while others took a more measured view of causes.

Before the 1980 publication of the asteroid impact theory of the extinction for nonavian dinosaurs, interest in Mesozoic mammals in general and Late Cretaceous mammals in particular was limited to paleontologists and a few diehard mammalogists. After publication of the Alvarez hypothesis there was a much keener interest in all things fossil just before and just after the K/T boundary. It would be too sweeping to say that the appearance of the impact theory caused a renaissance in the study of Cretaceous mammals, but unquestionably it was an important factor in taking them off dusty museum shelves. This is because not just Cretaceous mammalian paleontologists were interested in these organisms. Now extinction theorists—those interested in how this event related to the radiation of mammals—and members of the budding field of molecular systematics began to weigh in on the importance of the K/T extinctions and what followed. Stephen J. Gould famously noted at various times and in various venues that if nonavian dinosaurs had not become extinct we would not be here arguing about the cause of their extinction.

In the introductory chapter to a 2004 compilation titled *Mammals from the Age of Dinosaurs,* the authors note "the number of known Mesozoic genera increased to 283 by the year 2000. New Mesozoic mammals discovered in the past 20 years are one and one-half times more than the total of those known to science from the previous 200 years combined."[23] That is an impressive number. While much of this new work is noteworthy, I will here comment more fully only on the work since 1980 that has impinged most directly on the question of where and when modern groups of eutherian mammals, the group to which we belong, arose. As we see in Chapter 3, this wealth of new information has allowed us to begin to place the origins of eutherian groups in a better phylogenetic and biogeographic context.

One of the most enduring projects since 1980 has been the joint Mongolian (Mongolian Academy of Sciences) and American (American Museum of Natural History) Gobi expeditions, which, since 1990, have recovered immense troves of Cretaceous vertebrates, arguably including the richest and most complete specimens of Cretaceous mammals. Like the Poles before them, these teams have begun to describe these mammals in considerable detail. While a variety of mammals has been found and described, two quite recent contributions are the monographic studies of the eutherian mammal discussed earlier, *Zalambdalestes*,[24] and a newcomer to the Cretaceous pantheon of Cretaceous eutherian mammals, *Maelestes*.[25]

Elsewhere in Asia, and as noted above, in Uzbekistan, work has continued since the 1990s, resulting in a number of papers dealing with the dominant mammalian group in the faunas, the eutherians.[26] There may be as many as 13 eutherian species in the major sites in Uzbekistan. Much rarer Late Cretaceous eutherians have also been reported from India,[27] from Japan,[28] and in Europe from Spain and France.[29]

In Asia and specifically in China we are also seeing an explosion of often nearly complete but sometimes quite squashed specimens of Jurassic and Cretaceous mammaliaform mammals. Because of the completeness of the spec-imens we have for the first time clear evidence of definite aquatic and gliding forms of Mesozoic mammals.[30] The quite complete but very flattened Early Cretaceous Chinese *Eomaia* and *Sinodelphys* are claimed to be the earliest eutherian and metatherian, respectively.[31]

Extensive work has also continued and expanded since 1980 in the Western Interior of North America. Unlike in Asia, where eutherians are found in the fossil record from at least 105 millions years ago, they do not show up in the North American fossil record until the Late Cretaceous, about 75 to 80 million years ago, except for a single lower jaw from southern Montana dating from about 100+ million years ago.[32] Eutherian-producing Late Cretaceous North American sites cover a considerable geographic range, from Alberta and Saskatchewan in Canada, south into Montana, the Dakotas, Wyoming, and Utah, and further to the south in New Mexico, and in Texas.[33] There are even Late Cretaceous eutherians known from Baja California.[34] If all Late Cretaceous North American mammal-bearing sites were counted and not just those yielding eutherians, the list would be notably longer.

3

In Search of Our Most Ancient Eutherian Ancestors

MAMMALS LIVING TODAY COMPRISE three major branches, or clades. Including their fossil members, these clades are Prototheria, Metatheria, and Eutheria. Prototheria was probably never rich in species and today includes only five species of the crown clade, Monotremata[1]—the echidnas and platypus of Australasia.[2] Metatheria is decidedly more species rich than Monotremata and was even more so in the past, especially in South America. Its crown clade is Marsupialia, which has some 237 species in Australia, 94 species in Central and South America, and only 1 in the United States (the opossum). Eutheria has dominated in terms of numbers of species since at least shortly after the K/T boundary. Today the eutherian crown clade, Placentalia, with 5,080 species, is surpassed by marsupials in numbers of species only in Australia. Metatheria and Eutheria, and their crown clades, Marsupialia and Placentalia, respectively, are within the more inclusive clade Theria, which is Greek for *beast*. They share a number of features, notably their dentitions, mode of milk extrusion through nipples or teats, and reproductive systems, that shows they shared a common ancestor not shared with Prototheria (including Monotremata) (see fig. 3.1).

In Chapter 2, I examined the history of the discovery and study of Mesozoic mammals. In Chapter 6, I compare the molecular and morphological evidence regarding the geographic history and the timing of the origin of the crown clades—Monotremata, Marsupialia, and Placentalia—with considerably more emphasis on the taxonomically dominant placentals. First, however, I explore the record of Cretaceous eutherian mammals, emphasizing those that have been implicated in the origin of living groups of placental mammals. The question of the relationship between stem Eutheria and its crown clade, Placentalia, has been studied far more closely by more researchers, hence in this chapter I discuss almost exclusively those Cretaceous eutherian groups that have been implicated in the origin of Placentalia and its included orders. In addition, much of my work lies with eutherians, so I start with groups my colleagues and I have worked on from the Middle Asian country of Uzbekistan—zhelestids, zalambdalestids, asioryctitheres, and the obscure *Paranyctoides*. I also discuss some North American Late Cretaceous eutherians—cimolestids and *Gypsonictops*.

NO PATTER OF LITTLE HOOFS

Time has obscured the exact sequence of events, but during the early 1990s Lev Nessov, of Leningrad State University in Russia, began to work with Zofia Kielan-Jaworowaska, then of the University of Oslo, on the phylogenetic analysis of various mammals from the Kyzylkum Desert of Uzbekistan. Kyzylkum is from the Kazakh

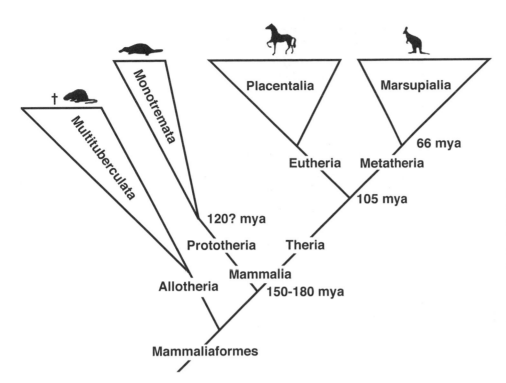

Fig. 3.1. Cladogram of Mammalia-formes. Thickness of triangles of crown groups do not represent relative species richness of that group. Note that the position of Multituberculata is in dispute. I accept the view based mostly on cranial and dental evidence that this clade is outside Mammalia. As discussed in the text, the split between Eutheria and Metatheria could have happened as long as 125 million years ago, based upon some as yet not well-illustrated or well-described specimens from China.

language and means "red sand." At 300,000 km², Kyzylkum is one of the largest and most remote deserts in the world, ranking somewhere around eleventh in size.

One of Nessov's groups of mammals that appeared to be new to science but was common at Dzharakuduk in Uzbekistan was the so-called zhelestids. The place name *Dzharakuduk* translates from Kazakh as "well by the escarpment," referring to an artesian well at the base of the 100-meter-high hills and cliffs that dominate the landscape. Zhelestidae is an extinct family of mammals named for one of the included species, *Zhelestes temirkazyk*. This is a species that Nessov had named in 1985. He had a propensity to provide names that, when spoken, rolled off the tongue. *Zhelestes* is derived from the Kazakh and Greek languages—*zhel*, which is Kazakh for "wind," and the Greek *lestes*, "robber." The species name *temirkazyk* derives from the Kazakh word for the North Star, which is "iron pin."

Kielan-Jaworowaska and Nessov approached me about reviewing a manuscript they were preparing on the zhelestids because the zhelestids' teeth bore an uncanny resemblance to the teeth of the earliest possible relatives of the hoofed mammals (e.g., horses, cows, deer, etc.) known as condylarths, or archaic ungulates. Condylarths are known with certainty only from the Cenozoic, although Dick Fox, from the University of Alberta, has reported Cretaceous species from sites in western Canada.[3] In my work on Late Cretaceous and early Tertiary mammals in North America, I had worked extensively on the early condylarths, dating from about 66 million years ago, and thus might provide comments on the zhelestid manuscript. *Condylarthra*, which literally means "knuckle joint" (Latin, from Greek), was named in 1881 by

the well-known nineteenth-century paleontologist Edward Drinker Cope, presumably because the anklebone known as the astragalus resembled a rounded knuckle.[4] Cope thought that condylarths included the ancestry of what we today collectively call ungulates—the odd-toed perissodactyls, such as horses and rhinos, and the much more speciose even-toed artiodactyls, such as cows, deer, giraffe, and their much-changed relatives, the cetaceans, or whales.

I generally agree with Cope's assessment, but the problem has been trying to tie specific condylarths to the ancestry of any living ungulate group. In fact, the other name for the group, *archaic ungulates,* is somewhat of a misnomer because, except for a few small hooves on a few condylarths, the species in this group do not have hooves, which the name *ungulate* implies. So we have this very diverse early Tertiary condylarth group whose descendents and ancestry remain unclear. This quite amazing diversity of early Tertiary condylarths indicates something of the process and pattern of mammalian evolution after the extinction of nonavian dinosaurs (see Chap. 6).

Soon after asking me to comment on their zhelestid work, Nessov and Kielan-Jaworowaska graciously invited me to join them as an author on their study. I jumped at the opportunity. I had been fascinated as to why the archaic ungulates seemingly appeared in North America either just before or just after the end of the Cretaceous without having a clear relationship to other Late Cretaceous mammals in North America. The zhelestids might offer an answer.

In 1994 I visited Kielan-Jaworowaska in Oslo, Norway, where she was the director of the Paleontology Museum at the University of Oslo. We began the joint study of zheles-

Fig. 3.2. Lev Nessov in his home study, St. Petersburg, Russia, 1994.

tids in earnest at that time. I was soon en route to St. Petersburg to do similar research with Nessov, and then we were to go on to the Kyzylkum Desert, which was my first of nine visits. While Kielan-Jaworowaska had some casts of the zhelestids, Nessov had the real things at his home office in St. Petersburg (fig. 3.2). I could hardly wait to see and examine the real fossils. It is hard to describe my anticipation, but it cannot have been any less than the feelings of visiting Paris, Rome, or New York for the first time as a younger person, which I decidedly was not at that time.

Nessov and I pored over the zhelestid specimens day and night for at least a week. Here were two people with a common purpose, similar interests, and a similar academic track, but whose backgrounds could not have been more different. There is one particular way that this played itself out. I had been trained in a methodology of phylogenetic analysis known as phylogenetic systematics, or cladistics, which I had read about as an undergraduate in the early 1970s but whose significance I did not understand at the time. It was not until graduate school at the University of California at Berkeley a few years later that it began to make sense. Although this methodology had gone far beyond, and continues to go even further beyond, its originator, the German entomologist Willi Hennig, it had made little if any inroads into Soviet science. Some younger Russian colleagues suggested to me that part of the stumbling block was that the methodology was the work of a German scientist, which made the older generation of Soviet scientists slow to accept it (shades of the past). My friend and colleague Alexander Averianov disputes this interpretation. He points out that Russian science arose from German science and that most older Russian scientists

knew German rather than English. According to him, the problem is that Hennig's work itself was quite outside the German scientific tradition. The main reason it did not find adherents in Russia was because of its elaborate methodology of systematics, according to which cladistics looks quite simple. Whatever the case, it was a methodology whose time had not yet come in Russia.

According to my colleague Alexander Averianov, who was a student of Nessov's at the time, the seminar on phylogenetic systematics I offered while a Fulbright Fellow at St. Petersburg University in the fall of 1996 it was the first such course offered in Russia. It should be no surprise, then, that Nessov and I spent much time in 1994 debating the appropriate methodology to test the evolutionary affinities of zhelestids. Eventually, he convinced himself that cladistics was the proper approach. Kielan-Jaworowaska, while still somewhat skeptical, had had much more interchange with western colleagues and was thus no stranger to cladistics.

In 1994 and into 1995 Nessov, Kielan-Jaworowaska, and I continued to compile a matrix of characters for zhelestids, other Cretaceous mammals, archaic ungulates, and early members of a few other groups of placental mammals. In cladistics, building a character matrix is the first step after carefully analyzing the anatomy of the species. If the anatomy is not carefully studied, a matrix and the resulting cladistic analysis can provide false or misleading results. Thus there are no short cuts in a thorough study of the anatomy of species under consideration. Even if properly done, most cladistic analyses do not often yield a single, definitive answer concerning the relationships of the species being studied. Apart from any problems with the methodology, evolution is not a nice, neat linear process but has a number of twists and turns: species can evolve convergent anatomy, meaning the evolution of their similar features is the result of similar environments rather than a close relationship; they can go through apparent reversals in their evolutionary trajectory; and they evolve at faster or slower rates. All these complications make the building of the tree of life a difficult task. The advent of molecular techniques in phylogenetic studies has brought a whole new realm of possibilities for unraveling evolutionary relationships, but it has brought its own set of problems as well. Whether it is molecules or anatomy, a cladistic analysis is only as good as the characters and taxon sampling it uses, and for all such studies, we are trying to group species together based on characters that evolved in their most recent common answer—what we call shared derived characters.[5]

As part of the zhelestid study, we also prepared detailed descriptions and figures with the intention of making them part of a major paper to be included in a book on Asia as the cradle of many placental mammals groups that was being planned by colleagues at the Carnegie Museum of Natural History. If the cladistic analysis did indicate a zhelestid–archaic ungulate link, a shorter paper in one of the higher profile journals such as *Nature* or *Science* might be warranted. The results did show this link quite strongly and a shorter paper was begun.

Tragically, Nessov died in the fall of 1995, before the paper was written. His non-Russian as well as his Russian colleagues mourned his loss as a colleague and a friend. I completed the shorter manuscript in 1996, after Nessov's death, and submitted it to *Nature*, a high-profile journal that was favorably disposed to publish results for paleontology. Surprisingly, they rejected it without review. I then sent it to *Science*, which at that time was much less inclined to publish such papers. To my delighted surprise, it was quickly accepted.

Like most everyone else, scientists like to see their work appreciated. Even though departmental chairs and deans too often measure academic success in terms of dollars rather than quality science, professional colleagues fortunately recognize one another's research, whether in the form of praise or criticism. It has been argued that science is unique in that it learns from its mistakes and moves on to new hypotheses supported by new evidence. While true, because people do science, it is sometimes hard to cast aside a favorite hypothesis in lieu of a new one. Change in science is inevitable; scientists either shake out the old hypotheses as new ones come along or die with the old ones, making room for the new.

While the idea that zhelestids were the most ancient relatives of ungulates or living hoofed mammals was certainly not a scientific revolution, it did have ramifications beyond paleontology. For example, a paleontological colleague of mine at Brown University, Christine Janis, indicated that some geological associates thought the argument was being forwarded that there were rhinos in the Cretaceous! With her wicked English sense of humor, Christine may have been pulling my leg somewhat, but it probably was difficult for some non-systematists to understand that we were arguing that the zhelestids were the ancient relatives of ungulates, not the ungulates themselves. In the jargon of cladistics, we were arguing that zhelestids were the sister group of ungulates. Part of this misunderstanding occurred because I was, and still am, a vocal skeptic of an asteroid impact as the sole cause of the terminal Cretaceous extinctions. Thus, some colleagues mistakenly saw this as an attempt to argue that some lineages of mammals went through the Cretaceous/Tertiary boundary unscathed. Such a charge was of course absurd.

All species of plants and animals have ancestors, even if it is extremely difficult to detect them in the fossil record. What we were arguing was that one of the ancestral lineages of a modern group of mammals had been recognized in the Late Cretaceous, not the lineage itself. Ramifications of this are detailed in Chapter 6; here I continue with the more specific issue of whether zhelestids are Late Cretaceous representatives of the crown group Placentalia or are stem eutherians outside of this crown group.

Although there had been suggestions in the past that specific Cretaceous eutherians may have given rise to specific placental mammalian orders, the publication in 1996 in *Science* and the larger study published in the *Carnegie Bulletin* in 1998 explicitly made zhelestids the nearest sister group to ungulates. Molecular systematists were uniformly pleased that at least possible placental mammals were now represented by several fossil species in the Late Cretaceous. The paleontological community was more divided. Studies have continued since those by Nessov, Kielan-Jaworowaska, and me. Hundreds of additional zhelestid jaws, teeth, postcrania, and ear regions have been found in the field seasons subsequent to 1994 by our nine URBAC expeditions,[6] spanning 1997 through 2006.

At the present time Nessov's former student and now head of the mammal lab in St. Petersburg, Alexander Averianov, and I have been able to recognize five different zhelestids at the 90-million-year-old Dzharakuduk site, as well as three more species at 95-million-year-old sites in Uzbekistan. At Dzharakuduk, going from smallest to largest, these are *Aspanlestes aptap*, *Parazhelestes mynbulakensis*, *Zhelestes temirkazyk*, *Parazhelestes robustus*, and *Eoungulatum kudukensis*. The names mean, respectively, "sky thief in the full blaze of the sun," "near the wind thief of the thousand springs," "wind thief of the north star," "the large form near the wind thief," and "dawn hoofed beast from the well." Our taxonomic studies have been able to reduce the number of species named by Nessov, but more importantly, as discussed below, additional studies are revealing information on the ears and skeletal elements of these creatures.

Most people, at least in North America, are familiar with raccoons. Raccoons are true carnivores in that they belong to the same order to which dogs and cats belong—Carnivora. Unlike dogs and cats, however, raccoons and most of their relatives in the procyonid family are more fully omnivorous or even frugivorous. This is borne out by their teeth, which have relatively low, rounded cusps. The canines are retained as quite large teeth. Superficially, this describes at least the anterior part of a large zhelestid skull. The largest zhelestid, *Eoungulatum*, approached the size of a small raccoon or even more closely that of the raccoon's smaller relative, the ringtail, *Bassariscus*, found in the American southwest and Mexico. Figure 3.3 is a series of reconstructions showing what a larger zhelestid may have looked like.

Going downward in size, the smallest zhelestid, *Aspanlestes*, was about the size of a rat, although its teeth were more anatomically similar to those of the ringtail. The ongoing postcranial research by Eric Sargis and his graduate student at Yale University, Stephen Chester,[7] tends to point to zhelestids as being terrestrial animals, as indicated in the reconstruction in figure 3.3. Some of the larger postcranial remains suggest that, like the raccoon or possibly more so the opossum, zhelestids were quite general in their postcranial anatomy, while others show specializations not seen until later placentals. Even if zhelestids were very early kin of ungulates and their relatives, there is no evidence that they possessed little hoofs. More likely, they primitively bore claws on most or all of their digits. These speculations go along with the better-known aspects of the teeth and skull that identify them as more generalized omnivores.

With the discovery and identification of the best-known zhelestids from the Kyzylkum Desert, a totally new clade of Cretaceous eutherian mammals had been recognized in the past 10 years. Zhelestids seemed to be showing up nearly everywhere in the Late Cretaceous. Some of the species were previously known, but with the recognition of the zhelestids, these species now had a place to hang their phylogenetic hat. In Spain and France, Emmanuel Gheerbrant, of the Natural History Museum in Paris, has described a separate lineage of European zhelestids.[8] In North America zhelestids include various Late Cretaceous taxa, such as *Avitotherium*,[9] closing the biogeographic gap across the Bering Strait cited earlier. A very fragmentary tooth from the Late Cretaceous of Madagascar originally thought to be a marsupial may well be a zhelestid.[10] Fragmentary material of a zhelestid has even been described from Japan.[11] The once obscure zhelestids now appear to have been one of the most widespread clades of Cretaceous mammal in the world.

Although some aspects of zhelestid anatomy do bear a striking resemblance to archaic ungulates, there are difficulties. The earliest known archaic ungulates in the Paleocene of North America and the earliest zhelestids in the Late Cretaceous of Uzbekistan are separated by 20 to 25 million years and are a continent apart. The geographic distance was not that important because it was clear that, since at least the time of the nonavian dinosaurs, animals had crossed back and forth over the Bering Sea land bridge during the times it rose above the waves. The first people to arrive in North America almost certainly followed this route, whether by boat or on foot. And, of course, although not known from the earliest Tertiary of Asia, archaic ungulates may well have been in Asia at this time as well.

The real problem lay in the considerable difference in age between zhelestids and archaic ungulates. When I first reviewed Nessov and Kielan-Jaworowaska's manuscript I found that I agreed with them that the molars of the zhelestids were lower, with more rounded bumps (known as cusps), compared to those of most Cretaceous contemporaries, such as asioryctitheres, signaling that the zhelestids were shifting their diet away from slicing and dicing insects and small vertebrates and going toward a more omnivorous diet, one that included both animals and plants (fig. 3.4). Such teeth characterized the earliest archaic ungulates as well. Thus, the zhelestid molars bore an uncanny resemblance to those of the earliest archaic ungulates. Even more exaggerated forms of these teeth are found in three of the most truly omnivorous mammals alive today—bears, pigs, and humans—all of which evolved such dentitions from different ancestors, by the process known as convergent evolution alluded to earlier (fig. 3.5).

In addition to the lowering of their molar crowns, zhelestids had a jaw, or mandibular condyle, that was set above the level of the tooth row. In humans, the mandibular condyle is part of what your friendly neighborhood dentist calls the TMJ, or temporomandibular joint, the joint connecting your lower jaw to your skull. In the ancestral mammalian condition, the mandibular condyle is near the same level as the top of the cheek teeth—the teeth behind the canines. What this means functionally is that with the ancestral condition, the back teeth contact each other first, with progressive contact moving forward in the jaw, as in scissors. There are exceptions; notably, the generally much longer canines pass each other early in the bite sequence. This sort of jaw arrangement is much as it is today in cats and dogs, which tend more to slice and then bolt, rather than chew their food (fig. 3.6).

In humans, horses, cows, and so forth, and to a lesser extent in zhelestids, with the mandibular condyle above the top of the cheek tooth row, the cheek teeth contact each other at about the same time. This is similar to diagonal pliers, in which the joint is set in a different plane than the jaws of the pliers. With this arrangement in humans, horses, cows, and zhelestids, food is chewed on one side of the jaw, as all teeth on that side contact each other at nearly the same time. With this sort of jaw arrangement and lower molar crowns, zhelestids had shifted away from the slicing and dicing seen in their ancestors to more crushing and chewing of food, which came to include plant material. This is the first time in eutherian evolution that this kind of feeding and diet is known. Recall that eutherians are all living mammals that share a more recent common ancestor with humans than they do with metatherians, such as the opossum. Also recall that placental mammals are the crown, or living members, of Eutheria.

Fig. 3.3. Study of the basal eutherian *Zhelestes*. Skull is about 6 cm. Reconstructions by Maria Gonzalez.

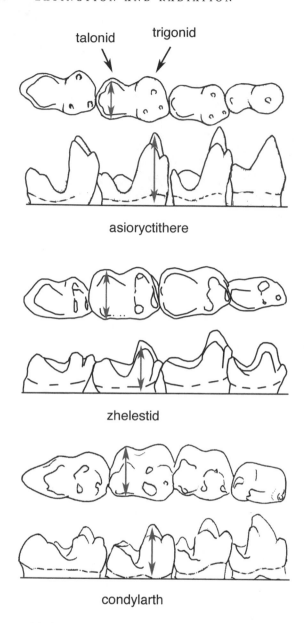

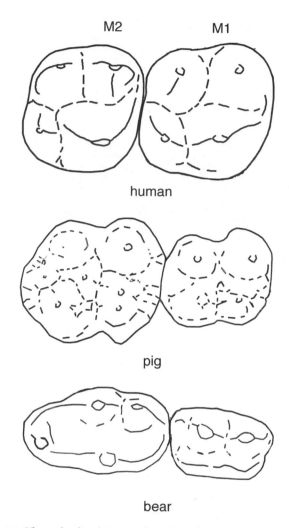

Fig. 3.5. The occlusal, or biting, surface of the first and second upper molars of a human, a pig, and a bear. Such teeth are characteristic of omnivorous mammals. They are rectangular or square in outline through the addition of cusps. The cusps are low and the surface is often crevassed. Not to scale.

Fig. 3.4. The last premolar and three molars from the left dentaries of an asioryctithere, a zhelestid, and a condylarth. The upper set for each is the occlusal, or biting surface, and the lower for each is the lingual, or tongue, side. Note the wider talonids and the lower trigonids in both zhelestids and condylarths, indicating a shift away from slicing and piercing and toward crushing and grinding. Not to scale.

Although these aspects of the zhelestid dentition and jaw just discussed are quite derived as in many much later placentals, zhelestids also retained quite a few ancestral or primitive eutherian aspects, not only in their dentition and jaw, but also in their ears and limbs. A major question this pattern raises is whether zhelestids, with this mixture of ancestral and derived characters, were near the base of the radiation of eutherian mammals or shared a closer relationship with some extant placentals (fig. 3.7).

One of the more ancestral eutherian features of zhelestids can be found in the greater number of teeth that they possess. Mammals almost always have the same number of teeth on the right and left sides of their mouth. This is generally true of the upper and lower teeth as well, but there are many exceptions. The dental formula refers to the number of upper and lower teeth on one side of the mouth. Except for unusual cases of supernumerary, or extra, teeth, especially in some aquatic mammals, all placental mammals today have on each side of their jaw at most three upper and lower incisors, one upper and lower canine, four upper and lower premolars, and three upper and lower molars. This results in what is known as a dental formula of I3:C1:P4:M3/i3:c1:p4:m3, or expressed more simply, 3:1:4:3 / 3:1:4:3. As noted, this illustrates the maximum number of permanent teeth borne on upper and lower jaws on one side of the head. Modern pigs retain this formula and dogs almost do, except for one upper molar less on each side.

This is also the dental formula for the earliest known ar-

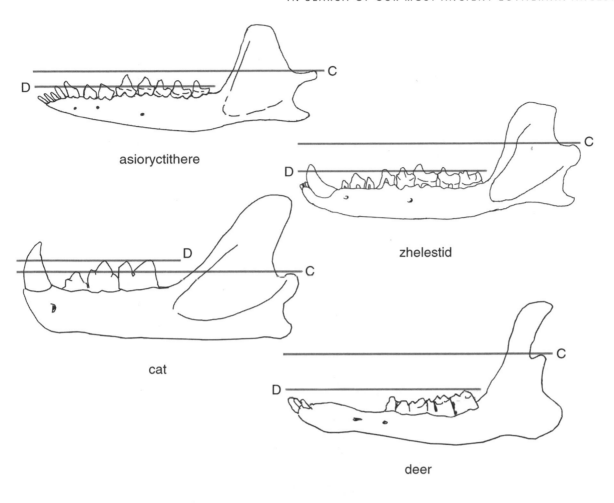

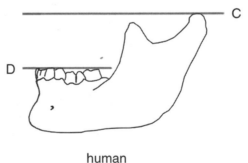

Fig. 3.6. The dentary, or left lower jaw, of the two basal eutherians, an asioryctithere and a zhelestid, compared to that of a cat, a deer, and a human. Ancestrally, the condyle (C) at the back of the dentary is almost at the level of the teeth (D), as in an asioryctithere, but in some Cretaceous mammals, such as zhelestids, the condyle is higher. Carnivores such as cats are derived in lowering the condyle, whereas herbivores such as deer and omnivores such as humans are derived in having the condyle even higher. See text for discussion. Not to scale.

chaic ungulates, but not for most extant hoofed placentals or even humans. Humans (as well as all apes and Old World monkeys) have a dental formula of 2:1:2:3 / 2:1:2:3. Thus, compared to the greatest number of teeth normally known in living mammals (or early archaic ungulates), humans have lost one upper and lower incisor and two upper and lower premolars on each side of the jaw. It is the last, or third, molars, the so-called wisdom teeth, that are most often pulled by the dentist when they fail to erupt properly. In some human populations, especially those that have had agriculture for thousands of years, the wisdom teeth may even fail to appear, a process known as agenesis of the third molar. People without third molars did not run the risk of an impacted tooth and

hence had a very slight selective advantage over those that had this tooth. Thus individuals with a third molar might die younger from dental problems, leaving no or fewer offspring. This is the evolutionary process of natural selection operating on human dentition.[12]

Compared to either living placental mammals or the earliest archaic ungulates, zhelestids had four, not three, lower incisors and five, not four, upper and lower premolars on each side of the jaw. We do not have well-preserved parts of the snouts of these mammals, but they may also have had five rather than three upper incisors on either side of the upper jaw. This is likely because some of the eutherian mammals from the 75-million-year-old beds in Mongolia, which are 15

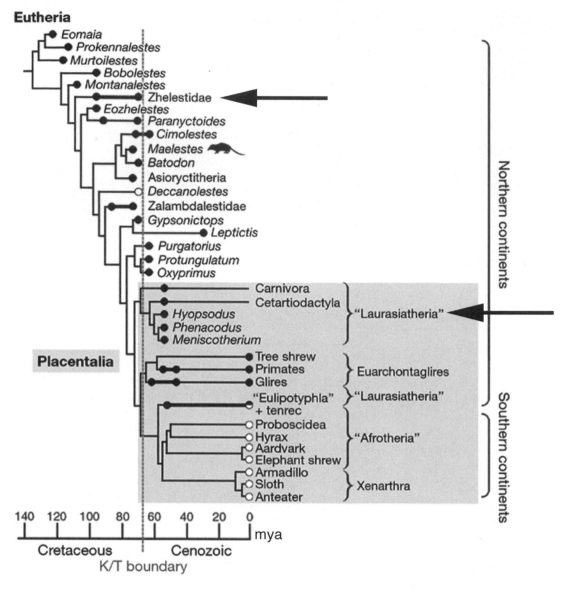

Fig. 3.7. A cladogram of fossil and recent eutherians by Wible et al. 2007. Note that the authors found no crown eutherians (placentals) before the K/T boundary. All Cretaceous and some early Tertiary taxa are basal to placentals. In our original analyses, zhelestids grouped with laurasiatheres (lower arrow), whereas in their analysis, zhelestids are near the base of Eutheria.

million years younger than the Dzharakuduk zhelestids, also had five upper incisors (and four lower incisors).[13] Members of the extant opossum family also have five upper incisors and four lower incisors. This suggests that the common ancestor of metatherians (including marsupials) and eutherians (including placentals) had five upper and four lower incisors on either side of the jaw. Also interestingly, some zhelestids, like some other early eutherians, have a canine with two roots, while others had only one canine root. With very few exceptions, mammals today, including humans, have only one root for their canine.[14] Having two roots may even be ancestral in mammals, but what the zhelestids show is that this character was still quite plastic some 90 million years ago, that is, selection or possibly chance had not yet settled on one canine root as the overwhelmingly common condition in mammals.

Other parts of the anatomy of zhelestids appeared to be more like those of basal eutherians than like those of extinct and extant placental ungulates. The problem was that originally none of these other fossil parts we had recovered could definitely be identified as a zhelestid because they were found as isolated elements. One of these bones is called the *petrosal*. The origin of the name *petrosal* comes from the Greek word *petra*, meaning "rock," as in the name Peter. Hence the biblical admonishment from Jesus to Peter, "Thou art Peter, and upon this rock I will build my church."[15] Because the petrosal is solid and rocklike, it is often preserved even when other bones have long ago eroded away. The petrosal houses part of the middle ear and has a bony covering for the coiled inner ear, which houses the hair cells that transmit auditory nerve impulses to the brain. This coiled inner ear is called the *cochlea,* a word of Greek origin that means "spiraled snail shell"—an appropriate name. It also houses the three semicircular canals, which are set at right angles to each other, helping to orient us in space. They provide our equilibrium. The petrosal is very useful for phylogenetic studies because it is a crossroads of sorts for a number of small arteries and veins, as well as nerves that often leave their impressions on the bone.

One of my then graduate students, Eric Ekdale, included the study of these petrosals as part of his thesis research. As noted, he had to assign the petrosals to various mammalian groups based on size and frequency compared to identified dental remains. Later we found petrosals associated with teeth showing that Eric's identifications had been correct. What he found were petrosals that, while definitely eutherian, retained some primitive characters of the venous circulation and openings, called foramina, that were similar to those of the common ancestor of metatherians (including marsupials) and eutherians (including placentals). He also found, by using high resolution x-ray computed tomography (CT), that the cochlea in these zhelestid petrosals coiled only slightly more than one turn.[16] Many placental mammals have a number of turns of the cochlea, although some have fewer turns, for example, zhelestids and earlier mammals. While not definitive, the fewer turns in the zhelestid cochlea was one more indication of this group's early split within eutherian mammals rather than their being an ancient relative of archaic ungulates.

As noted above, colleagues elsewhere studied isolated skeletal elements from our sites in Uzbekistan.[17] These colleagues, who specialize in postcrania (skeletal elements), studied numerous anklebones, known as the calcaneus and astragalus, forelimb elements, called the humerus and ulna, and one hind limb element, identified as the femur. The work is ongoing, but two general patterns have emerged, one functional and one phylogenetic. The functional pattern is the least controversial. The postcranial elements that have been associated with zhelestid dentitions and petrosals indicate, based on size and abundance, that zhelestids were more terrestrial rather than being primarily arboreal, or tree dwelling.

Issues, however, arose regarding the phylogenetic affinities of some of the postcranial elements. My colleague Alexander Averianov and I had established, based upon dental remains, that there were as many as 13 species of eutherians at Dzharakuduk and only one species of metatherian. According to our colleagues, there were at least five metatherians, based on postcrania. There certainly may have been more metatherians from Dzharakuduk than the dental remains and petrosals indicated, or, less likely, Averianov and I could have misidentified some metatherians as eutherians. What has begun to emerge as more postcrania have been studied is that some of the characters of the postcrania that were thought to be metatherian in nature are in fact characters

ancestral to both metatherians and eutherians. Thus, as with aspects of the dentition and ears, zhelestids retain characters of their skeleton found in the common ancestor of metatherians and eutherians.

As the study of zhelestids has continued, it has become clear that they contain an interesting mix of primitive or ancestral therian (metatherians and eutherians) traits, along with clearly eutherian traits and more modern or derived placental traits. The condition of the petrosals, the limbs, and even the higher number of teeth all suggest that zhelestids branched near the base of Eutheria rather than being associated with the placental ungulates. On the other hand, the lower-crowned cheek teeth, along with a more advanced kind of jaw articulation, are found later in the earliest archaic ungulates. The question is, Do the ancestral characters indicate that zhelestids belong among the earliest eutherians or do the derived traits signal a close relationship with archaic ungulates, which are thought to be early placentals?

The retention of primitive characters in any species of animal, not just basal eutherians such as zhelestids, is not an oddity. Think of our own bodies. We have binocular, color vision and five toes on hands and feet. Binocular, color vision is shared with all other apes and monkeys, whereas five toes on hands and feet harks back to a much more ancient ancestor, who lived not long after we departed the water for good. All species are a mixture of these primitive and advanced traits, and thus teasing them apart can be a tricky affair.

Whereas there is little doubt or disagreement as to the importance of zhelestids as a diverse and important early radiation of eutherian mammals, their possible relationships to modern placental ungulates remained unclear. Studies that included zhelestids in their analysis affirmed this latter evolutionary linkage,[18] until an article appeared in *Nature* in 2007 by John Wible and his colleagues.[19] Ostensibly about the naming and description of the new eutherian mammal *Maelestes* from Mongolia, the true importance of the article was that, unlike our studies, which included all Late Cretaceous eutherian mammals but fewer placental mammals,[20] or studies by others that had many placental mammals but very few Late Cretaceous eutherians,[21] the study of Wible and colleagues had both ample numbers of Late Cretaceous eutherians and later placentals. Its most important conclusion was that no Late Cretaceous eutherians were shown to be a member of any placental group. The important ramifications of this study for the timing of the origin of placental mammals will be detailed in Chapter 6. What is key for consideration here is that zhelestids were found not to give rise to placental ungulates. Rather, they were shown to be a very early diverging branch of eutherians. The study may not turn out to be correct in all points, but it is for now the best such study we have. So zhelestids appear to retain many aspects of their dentition, ears, and limbs that are decidedly ancestral for eutherians. This means, if Wible and his colleagues are correct, that the lower-crowned, crushing sort of dentition of zhlestids evolved separately from the much later similar dentition of archaic ungulates.

NOT BUGS BUNNY'S UNCLE

Another mammal present at the 90-million-year-old Dzharakuduk sites was not represented by as many species as zhelestids but was nevertheless one of the most common mammals that we recovered. Named *Kulbeckia,* it is the earliest definite member of the family Zalambdalestidae (mentioned in Chap. 2). Unlike many of the names used for mammal species from Dzharakuduk, which are almost all derived from Kazakh, we are not sure of the etymology for this genus. *Kulbeckia* is named for Kulbeck Spring, which lies in the middle of a *takur* (fig. 3.8), which in Kazakh refers to a nearly horizontal salt-encrusted surface that in this instance stretches for 5 to 10 km. This becomes a shallow lake if there is enough rain, which suggests that *Kulbeckia* may have come from Kazakh *köl* for "lake" and *bek* for "big" or "strong," hence Big Lake Well. Another possibility is that *bek* also means "leader," so this may refer to the leader's lake or spring.

We have enough dental, cranial, ear regions, and bits of postcrania to show that *Kulbeckia* was rat-sized, terrestrial, quadrupedal, with some elongation of hind limbs for hopping, and may have fed on both insects and plants (fig. 3.9). Although our postcranial material is quite fragmentary, the portrait of *Kulbeckia's* locomotory habits are strongly enforced by the much more completely known postcrania of the larger, squirrel-sized *Zalambdalestes* and *Barunlestes* from the 15-million-year-younger sites in Mongolia.[22] These animals have been likened to elephant shrews (Macroscelididae), which today live in Africa. Elephant shrews are neither elephants nor shrews but derive their common name from their elongate proboscis and their smaller size, which incorrectly suggested a relationship to shrews. Both forelimbs and hind limbs are somewhat elongate, but the hind limbs more so, giving them a four-legged hop not unlike a rabbit's (fig. 3.9). This may have been especially true of the later and more derived *Zalambdalestes* and *Barunlestes,* but less so for *Kulbeckia.*

Unlike zhelestids, which are known with some confidence from Asia, Japan, Europe, and North America, zalambdalestids are known only from Asia. Interesting as well, zalambdalestids and zhelestids occur together in our Uzbek sites, but no zhelestids are found from Mongolia, where both *Zalambdalestes* and *Barunlestes* are known. One suspicion is that all sites that have zhelestids are lower coastal plain environments, while the Late Cretaceous of Mongolia was sometimes like the Gobi Desert today or sometimes a fluvial plain.

The most controversial aspect for zalambdalestids, as with zhelestids, is where they sit within the eutherian phylogeny. Are they basal eutherians or are they ancestral to some living placentals? While zhelestids were first recognized in the 1980s, the possible but probably now incorrect relationship of zhelestids to extant placentals was first argued only in the 1990s. This is not the case with zalambdalestids. They were thought to have some affinity with placentals from their first description in 1926 by Gregory and Simpson.[23] They did not suggest a relationship with any specific extant placentals, other than noting a similarity to the living but now endangered insectivore *Solenodon,* known from the islands of Cuba and Hispaniola, and a possible relationship to some extinct possible placentals.[24] In 1964, Leigh Van Valen was more specific, arguing that zalambdalestids were ancestral to lagomorphs—rabbits and their relatives.[25] In 1975 Malcolm McKenna carried this same theme further, placing zalambdalestids and

Fig. 3.8. Our camp at the Dzharakuduk escarpment, which is over 100 m high. The lower and upper beds are marine silts and shales. The lower-lying beds in the middle are the vertebrate-producing channel deposits. The flatter area in front is a *takur,* which in Kazakh refers to a nearly horizontal, salt-encrusted surface that in this instance stretches for 5 to 10 km. For scale, the larger yurt on the left is about 6 m across. Photo by Igor Danilov.

Fig. 3.9. Study of the basal eutherian *Kulbeckia*. Skull is about 4 cm. Reconstructions by Maria Gonzalez.

lagomorphs together in a then radically new classification of mammals.[26]

The most species-rich group of extant mammals, Rodentia, making up 42% of the living 5,400 species of mammals, soon came into—or returned to—the picture in the 1980s as possible rabbit relatives. Since the time of Linnaeus in the eighteenth century, rodents and rabbits had been associated "on and off." When "on," they were put together in a group called Glires (*glis* or *gliris,* Latin for "dormouse").

When we started working on all the new material of *Kulbeckia* that we were recovering from Dzharakuduk, Uzbekistan, we began to take an interest in the question of whether *Kulbeckia* and the other zalambdalestids had anything to do with the origin of Glires. *Kulbeckia* showed the earliest stages

of trends leading to the 15-million-year-younger zalambdalestids from Mongolia and then possibly to Glires.

If we start with a typical rodent and work backward in time, the similarities become apparent. Most of the really rodentlike features are in the skull and teeth. Rodents have a pair of upper and lower incisors that meet in order to gnaw anything in their way. Notably, the teeth are open-rooted, meaning they continue to grow throughout the life of the animal. If you deny a rodent something to chew it becomes quite agitated. Additionally, unlike in most mammal teeth, including our own, in which the enamel covers the entire crown, in the ever-growing incisors of rodents and rabbits the enamel is restricted to the front outside margin of the incisors. The remainder of the incisor is composed of dentine, the softer substance that underlies enamel on most mammalian teeth—again, including our own. What this means for the rodent is that as it gnaws something, the dentine exposed at the biting, or occlusal, surface wears faster than the enamel, creating a self-sharpening chisel. This is the key feature that led to the evolutionary success of rodents. Another measure of their success is that they are the only land-based mammal (not counting bats) that reached oceanic islands. Rodents reached the Galápagos Islands a few million years ago. If any nonflying mammal could accomplish this incredible feat, it would be rodents.

The other teeth in rodents, which are mostly three molars and often one premolar, vary widely from one group to another, but primitively they are low crowned. A large gap separates both these upper and lower cheek teeth from their respective incisors to the front (fig. 3.10). Such a gap is not uncommon in other mammals, but it is especially well developed in rodents because their snout is elongated, and all the more posterior incisors, canine, and almost all premolars were lost near the origin of the group. The gap is known as a *diastema* (plural *diastemata*), a Latinized Greek word meaning "interval" or "to separate." It functionally separates the acquisition of food by the incisors from the chewing of food by the molars. The tongue moves the food from the front to the side, something we also do to some extent while chewing. When the incisors are in contact the molars are not, and vice versa. Rabbits and their relatives generally share the features just described for rodents, but in addition they have a second set of small upper incisors, which set immediately behind the larger pair of upper incisors, forming a stop for the lower incisors (fig. 3.10).

In going from *Kulbeckia* to the later zalambdalestids, the diastemata increased in size by the reduction and loss of teeth (premolars) between the incisors and cheek teeth (fig. 3.11), the molars became lower crowned and more distinctly triangular and elongated side to side, as in some very early rodents. More spectacularly, even in the early *Kulbeckia*, the pair of lower medial incisors are greatly enlarged, lean decidedly forward, have enamel greatly thickened on the lower

and outer sides, and the incisor is probably open rooted—all very rodentlike features. The snout, although narrowed and elongate, does not have the peculiar elongation found in Glires, and most tellingly and un-Glires-like, there are no enlarged upper incisors.

In the same phylogenetic study in 2001 in which we found that zhelestids were linked to archaic ungulates, we found zalambdalestids to be linked to Glires.[27] This meant that both relatives of Glires and some relatives of ungulates had existed over 90 million years ago in Asia. This had ramifications about the timing of the origin of placental mammals (detailed in Chap. 6). The same publication in *Nature* in 2007 by John Wible and his colleagues that included zhelestids also included zalambdalestids.[28] They found no zalambdalestid-Glires tie, which again argued that no Late Cretaceous eutherians were shown to be a member of any placental groups (see fig. 3.7).

SHARP IN TOOTH AND CLAW

A major Late Cretaceous clade, which I will refer to as Cimolesta, has in the past (but less so recently) been cast as ancestral both to extinct and to extant placentals. Following the already-cited 2007 paper by Wible and colleagues, Cimolesta includes two clades that generally fall within two biogeographic regions.[29]

First is Asioryctitheria, which includes 90- to 75-million-year-old species best known from Uzbekistan and Mongolia. These specimens can be quite well preserved, especially from Mongolia, sometimes including skulls and on occasion skeletal material. Asioryctitheres have been only indirectly implicated in the origin of placentals. From our work in Uzbekistan we have been able to recognize four species of these asioryctitheres: going from smallest to largest these are *Uchkudukodon nessovi*, *Daulestes kulbeckensis*, *D. inobservabilis*, and *Bulaklestes kezbe*. The names mean, respectively, "tooth from the three wells named for Nessov," "sand storm thief of the big lake," "unseen sand storm thief," and "spring thief of kezbe."[30] *Uchkudukodon* is certainly the smallest Mesozoic eutherian mammal, and among the smallest mammals that

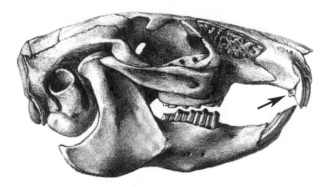
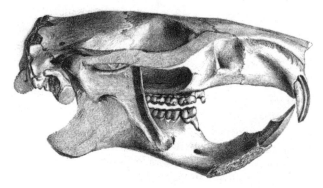

Fig. 3.10. Palatal (occlusal) views of the front ends of a rabbit skull (*left*) and a rodent skull (*right*) showing the second set of small incisors in the rabbit.

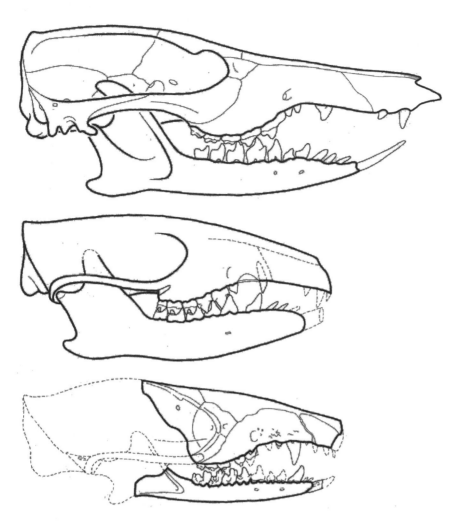

Fig 3.11. Skull reconstructions of the three better-known zalambdalestids—the 90-million-year-old *Kulbeckia (bottom)* and the 75-million-year-old *Barunlestes (middle)* and *Zalambdalestes (top)*. From Wible, Novacek, and Rougier 2004.

ever lived, with molars about 1 mm and a skull just over 1 cm (fig. 3.12).

The second clade, Cimolestidae, is essentially North American except for a new genus *Maelestes,* described by Wible and his colleagues in their 2007 paper. There had been some concern as to the monophyly of Cimolestidae, but one of my recent graduate students, Justin Strauss, was able to show quite convincingly that this family is monophyletic. It is members of Cimolestidae that have been most commonly argued to include the ancestry of some placentals. They have most notably been implicated in the ancestry of the extinct palaeoryctids, small insectivores with sharp cusped molars; taeniodonts, cat- to sheep-sized gnawing and burrowing forms; and creodonts, decidedly carnivorous forms from cat to leopard size, that were armed with multiple sets of sharp, shearing teeth. Creodonts are thought to be the sister group to Carnivora, the one extant order of placental mammals that has been argued to be a cimolestid descendent.

Although various authors had considered individually some of these suggested relationships, it was Lillegraven in 1969 who presented a phylogeny and explanation of why he thought all these relationships were possible.[31] One fea-

ture of cimolestids that impressed all workers was the very high crowned, thin-ridged, and sharp-cusped nature of their premolars and molars. They did and still do look like they could give rise to living carnivores, with their slicing carnassial teeth, but the 2007 analysis of John Wible and colleagues found them to be sister to asioryctitheres at a point more basal on the eutherian tree (see fig. 3.7).

THE SHREW NOT TAMED

Not all Late Cretaceous eutherian mammals can be neatly placed together into higher taxa, such as families, but rather seem to be orphans among their 90-million-year-old contemporaries. *Paranyctoides* fits this bill perfectly. This may in part be caused by the more fragmentary nature of the fossils attributable to this genus. The teeth, especially the molars, are lower crowned, even more so than in zhelestids, but they are not quite as bulbous, and the upper molars are somewhat squarer in outline.

What is quite remarkable about *Paranyctoides* is that it is the only Late Cretaceous eutherian mammalian genus recognized from two continents: North America and Asia. While not unknown for mammals, such recognition on two dif-

Fig. 3.12. Study of the basal eutherian *Uchkudukodon.* Skull is slightly larger than 1 cm. Reconstructions by Maria Gonzalez.

ferent continents is not that common. There are a few caveats to this assertion. First, some may think of Europe and Asia as separate continents, but for organisms, they constitute one big continent: Eurasia. Second, flying mammals (bats) and completely aquatic mammals (whales, sea cows, etc.) can be fairly excluded in this small comparison. Of the remaining 26 orders of living mammals, some 11 have genera (and in a few cases species) represented on more than one continent.[32] Some examples for mammalian genera (or species) found on more than one continent are in the following groups: artiodactyls (such as moose and reindeer), carnivores, cingulates (armadillos), didelphimorphs (opossums), lagomorphs (hares and pikas), perissodactyls (horses, rhino tapirs), pilosans (anteaters and sloths), primates, proboscideans (elephants), rodents, and soricomorphs (shrews). Except for a few genera of shrew, opossum, and primate that are quite small, all the

other genera (or species) found on more than one continent are medium to very large in size. When I write "small," I mean about the size of a common house mouse to that of a rat. The point of this exercise is to indicate that, although related small mammals may be found over wide areas, they seldom belong to the same genus or species. Thus, it is quite impressive that we have a mouse-sized genus of eutherian mammal, *Paranyctoides*, stretching from Middle Asia to western North America during the Late Cretaceous.

Paranyctoides was recognized and named by Dick Fox in 1979, with a second species added in 1984.[33] The material named in 1979 came from the Dinosaur Park Formation, the approximately 75-million-year-old dinosaur-bearing formation discussed in Chapter 1. Other species have been recognized from slightly older and younger beds possibly as far south as New Mexico.

When described, it was thought to belong to the extant placental order Lipotyphla, which includes modern shrews and hedgehogs, specifically in the extinct family Nyctitheriidae, otherwise known only after the K/T boundary. The name *Paranyctoides*, meaning "near nyctitheres," was meant to indicate this relationship.

Lev Nessov visited Dick Fox and in 1993 named *Paranyctoides aralensis*, after the Aral Sea.[34] He had been taken by the great similarity of the specimens from Canada and Uzbekistan. Alexander Averianov and I added new material of *Paranyctoides* and published a revision in 2001, which accepted Nessov's earlier assessment that *Paranyctoides* was known from both continents.[35] In 2007, my young Russian colleague Pavel Skutchas found a lower jaw with two well-preserved molars along with the alveoli or holes for many more teeth, which clearly represents a new species of either *Paranyctoides* or of a zhelstid. Pavel, Alexander, and I were screen-washing new sites in a remote desert area of Kazakhstan, about 100 km from the famous Soviet and now Russian Baikonur Cosmodrome, as part of an expedition run by our colleagues Gareth Dyke and Dmitri Malakhov. We were excited because this was the first clearly identifiable Late Cretaceous mammal from what were estimated to be about 80-million-year-old beds, slightly younger than in Uzbekistan to the south.

Until recently, I remained skeptical that we really had the same genus of eutherian mammal some 10 million years and thousands of kilometers apart. I had not seen original material from North America to compare with our Asian material. Then a few years ago a colleague in Oklahoma, Rich Cifelli, showed me other specimens of *Paranyctoides* from his sites in Utah. I was convinced. Based upon what we know at present, the specimens do appear to belong to the same group, the genus *Paranyctoides*. If nothing else it was another bit of information suggesting faunal interchange of not just dinosaurs but also mammals across what is now the Bering Sea.

Dick Fox had argued that the lower cheek teeth of *Paranyctoides* were "closer than is any other known Mesozoic mammal to what would be expected in an ancestor for the following placental groups: Lipotyphla [insectivores], Tupaiidae [tree-shrews], the orders Primates, Dermoptera [so-called flying lemurs], and Chiroptera [bats], and all of the ungulate mammals."[36] Alexander and I were especially impressed by the resemblance to zhelestids; thus *Paranyctoides* could simply be a somewhat aberrant zhelestid. In our phylogenetic analyses it tended to clump with zhelestids. The *Nature* paper in 2007 by John Wible and his colleagues clumped *Paranyctoides* with one zhelestid very close to other zhelestids in their phylogenetic analysis. These authors thus found a pattern similar to ours, but again they did not group these Late Cretaceous species with post K/T boundary placentals but rather found them to be quite basal eutherians (see fig. 3.7).[37] Once again the most recent studies did not find Cretaceous placentals.

THE LAST BEST HOPE

The final Late Cretaceous group to consider is composed of one genus, *Gypsonictops*, usually with four species assigned. It is known from about 75 to 66 million years ago, only in North America. While the molar cusps of *Gypsonictops* are of moderate height, they are not as sharp ridged and cutting as in cimolestids. Also, the last upper and lower premolar have undergone a process of molarization, wherein there are additional cusps that make these teeth look more like true molars. This process of molarization has occurred countless times in the later placentals; for example, in horses all the cheek teeth, be they premolars or molars, look very much alike—all tall square blocks of complicated folds that efficiently grind grass. But the process of molarization is much less common in Late Cretaceous stem eutherians, so *Gypsonictop* stands somewhat alone.

Gypsonictops is often placed in the monogeneric family Gypsonictopidae, with a sister group relationship to the Tertiary family Leptictidae, whose members were about the size of small rabbits. This relationship is quite strongly supported in most studies, but the relationship of *Gypsonictops* to other eutherians is more problematic. *Gypsonictops* has been implicated as ancestral to various placental lineages, but not when a cladistic analysis is performed. Wible et al. 2007 found that *Gypsonictops* (with its sister leptictids) was the closest Late Cretaceous eutherian relative to post K/T boundary stem eutherians and placentals, followed next distantly by zalambdalestids (see fig. 3.7). If their analysis is further supported, *Gypsonictops* and its kin may yet prove to be near the base of the placental radiation.

4 Patterns of Extinction at the K/T Boundary

DESPITE WHAT ONE MIGHT think, nonavian dinosaurs were not the only vertebrates tromping around in the Mesozoic world. This may well be an obvious truth to many, but a surprising number of people think otherwise. To help drive home this misconception, I have for a number of years shown a variant of a figure (fig. 4.1) showing a sunglass-wearing *T. rex* ambling along a highway somewhere out west. It is the Hollywood vision of nonavian dinosaurs that helps to color the perceptions of the public and even in some cases scientists. In the latter category is the plethora of books about the K/T extinctions that make nonavian dinosaurs the poster child of extinction but offer very little information about the extinctions or, more importantly, about the nonavian dinosaurs' many other contemporaries.

I must admit to some hyping of nonavian dinosaurs as well, having given over Chapter 1 to relaying what we know of nonavian dinosaur extinction. To make some amends I now turn to what we know of the remainder of the terrestrial realm, both on land and in fresh water. Again, I must emphasize that the vertebrate record surrounding the K/T boundary is only known with any degree of confidence from the northern parts of the Western Interior of North America. Here I examine the fossil record of each major clade of vertebrates before and after the K/T boundary and also comment on the plant and freshwater invertebrate record. While I would prefer to work at the lowest possible taxonomic level—the species, at which evolutionary processes operate—this is not always possible given the vagaries of the fossil record. Accordingly, in most instances I employ the more neutral term *clade*, which, as used by most systematists, means an arguably monophyletic group of organisms or higher taxa. For example, *Panthera leo* is a species-level clade, while Felidae is a family-level clade. Both are quite demonstrably monophyletic, although at different taxonomic levels. The reason for using monophyletic groups, or clades, is that it reduces by at least one parameter the number of variables involved in examining the patterns of extinction and survival across the K/T boundary. More ecologically based clusters, such as guilds, by their very nature are more difficult to delineate and track. The utility of clades will become more obvious when I turn to specific cases, starting with one of the more interesting and problematic instances—Mammaliaformes.

MAMMALIAFORMES

Chapters 2 and 3 discussed Mammaliaformes and their extant clade, Mammalia, which in turn includes Prototheria, with living Monotremata (platypuses and echidnas) and Theria. Theria in turn includes Metatheria, with the living Marsupialia, and Eutheria, with its living Placentalia. Figure 3.1 serves as a refresher for these clades.

Fig. 4.1. Cartoon of *T. rex* as a Hollywood star.

In Chapter 6 I return to the radiation of Mammaliformes after the K/T boundary, but we must first examine who was around and who survived.

What can be stated with about as much confidence as one can muster regarding the fossil record is that the three major clades—Prototheria, Metatheria, and Eutheria—are known from Cretaceous deposits and all survived the K/T boundary to give rise to their extant counterparts—Monotremata, Marsupialia, and Placentalia, respectively. What we cannot say with any great confidence is whether any of these three extant clades is known for the Cretaceous. This is explored in Chapter 6.

Another major player, at least in the Western Interior, are the wholly extinct Multituberculata.[1] This major clade was not noted in previous chapters. It may have been present on all continents, but it is best known from northern continents, collectively called Laurasia. The Cretaceous through early Tertiary was its heyday, but it became extinct in the middle Tertiary. Multituberculates also unquestionably survived the K/T boundary. If we limit ourselves to the Western Interior, it is clear that Multituberculata, Metatheria, and Eutheria all survived. Prototheria also survived, but only in Gondwana, specifically, the continents of Australia and South America. It will not be considered further.

Earlier published accounts recorded 10 Cretaceous multituberculates, with five surviving the K/T boundary, six of six eutherians surviving, and one of 11 metatherians surviving, all from sites in northeastern Montana.[2] With minor alterations discussed below, the multituberculate and metatherian records are only slightly changed by more recent research. Those for eutherians are more problematic.

In this earlier treatment,[3] some of the species included in the eutherian genus *Cimolestes* were thought to give rise to extant placentals or near relatives. This hypothesis is no longer tenable based on the 2007 results of John Wible and his colleagues, namely, that there is no evidence of any extant

placental clades in the Cretaceous. This eliminates one species, *Cimolestes cereberoides,* as a possible Carnivora ancestor,
but this species may still be linked to the ancestry of the extinct clade Taeniodontia. In fact, this order has been reported
from the Cretaceous of Canada.[4] In addition, problematic
Cretaceous occurrences of archaic ungulates (condylarths)
have been reported in Canada, but more recently other rare
occurrences of these archaic ungulates have turned up in
northeastern and southeastern Montana,[5] demonstrating—
at least in the latter case—that the condylarth genus *Protungulatum* was present, if rarely, in Cretaceous deposits in the
Western Interior. This adds another eutherian recorded from
eastern Montana and a definite Cretaceous placental from
the Late Cretaceous.[6]

Greg Wilson has added greatly to our knowledge of mam-

maliforms as well as other vertebrates from the Late Cretaceous Hell Creek Formation in northeastern Montana. He
has updated systematics from my earlier work, added new
species, and most importantly provided a longer biostratigraphic section of the Hell Creek Formation, which gives
more depth to the study of vertebrates in this critical time interval. Figure 4.2 is part of figure 2 from Wilson's 2005 paper
synthesizing the current state of our knowledge for Hell
Creek mammaliforms. For the 11 multituberculates that he
recognizes, two disappear well before the K/T boundary, six
disappear either before or at the boundary, while four clades
continue into the Tertiary. For metatherians, 11 species disappear before or at the boundary, while one or two clades
continue into the Tertiary. As discussed above, the eutherians
present a somewhat more complex story depending upon

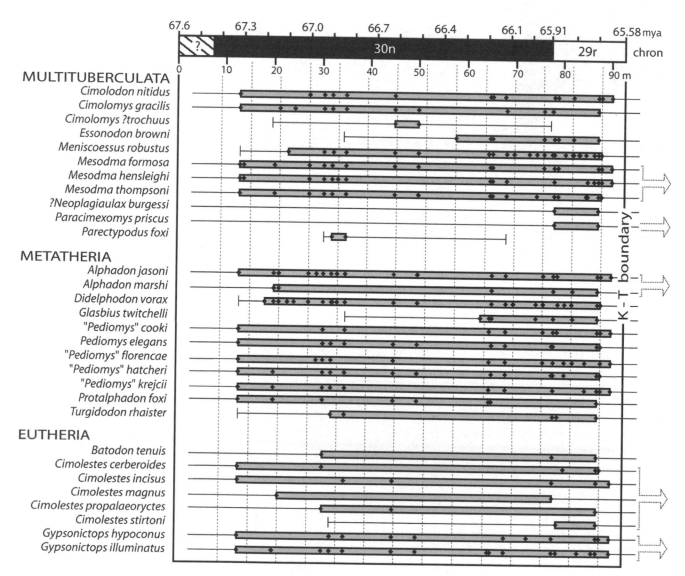

Fig. 4.2. Greg Wilson's 2005 synthesis of the biostratigraphic distribution of mammals in the Hell Creek Formation, eastern Montana, and their
probable fates at the K/T boundary. The terms *30n* and *29r* refer to numbered polarities of the Earth's magnetic field, with *n* meaning normal (as
today) and *r* meaning reversed.

how one interprets which clades do and which do not survive the K/T boundary. Wilson shows eight Hell Creek eutherians, with all but one surviving into the Tertiary. This would mean an 87.5% eutherian survival. A more conservative view is that we can only demonstrate with confidence that one species each of *Cimolestes* and *Gypsonictops* survived the K/T boundary, giving rise to known Tertiary relatives. This would mean only a 25% survival for eutherians.[7] As noted above, we can now more confidently argue that at least one species of Cretaceous condylarth survived into the Tertiary, which adds one more eutherian survivor.

Another complication of the mammal story across the K/T boundary is that if one accepts at least some of the molecular work, *all* orders of extant placental mammals are known before the K/T boundary. This subject is examined in detail in the final chapter, but suffice it to note here that the fossil record indicates that *none* of the orders of extant placental mammals are known before the K/T boundary. I accept the latter, well-supported interpretation of ordinal originations: there are no living orders of placental mammals known from the Cretaceous.

REPTILIA (ARCHOSAURIA)

Crocodilians, pterosaurs, and dinosaurs (including birds) together constitute the great clade known as Archosauria, or "ruling reptiles." The most basal living branch of this clade is the crocodilians (crocodiles, alligators, and relatives), followed by the extinct pterosaurs and dinosaurs as each other's nearest relative.

Crocodilia

Earlier tabulations for the Western Interior indicated five crocodilians in the latest Cretaceous.[8] These included three alligator relatives, one crocodile relative, and an unrelated species. All but one of alligator relatives, *Brachychampsa montana*, survived the K/T boundary.

There have been some taxonomic advances for this group.[9] The well-known species *Leidyosuchus sternbergei*, thought to be a crocodile or an alligator relative, is now assigned to a new genus, *Borealosuchus*, and appears earlier on the phylogenetic tree. Additionally, the species *Thoracosaursus neocesariensis* is now thought to be more closely related to the gavials among living crocodilians. Only a few fragments of this more commonly marine crocodilian have been recovered from the Hell Creek Formation. Two unnamed alligator relatives reported in the earlier tabulation noted above are known from both sides of the K/T boundary. *Brachychampsa montana* is still regarded as an alligator relative and, as noted above, did not make it through the K/T boundary.

Pterosauria

As with birds, I did not formally include pterosaurs in counts of extinction and survival at the K/T boundary in the Western Interior in my 1996 book on K/T extinctions because of their paucity in the fossil record. Since that time, the record has not increased greatly. We can say with some assurance, however, that at least one pterosaur is known from the Hell Creek Formation. For the present it is assigned to *Quetzalcoatlus*, the same genus that is better known from Texas, which, with a wingspan of perhaps up to 12 m, is the largest organism to fly. In turn, *Quetzalcoatlus* is included in Azhdarchidae, named by my friend and colleague Lev Nessov, for the pterosaur *Azhdarcho* from 90-million-year-old Late Cretaceous beds in the Kyzylkum Desert of Uzbekistan. Nessov named this pterosaur for the mythical dragon in Kazakh and Uzbek lore that is said to inhabit deep canyons, or *sais*, throughout the region. Not only does Azhdarchidae include the largest pterosaurs, but it may also have been the latest surviving family. It has been argued that birds may have been outcompeting pterosaurs by the end of the Cretaceous. All we can really say with some certainty is that the few remaining species of pterosaur were extinct at least by the K/T boundary. Whether they survived to the boundary is unknown.

Dinosauria (including birds)

Chapter 1 provided a detailed discussion of nonavian dinosaurs from the latest Cretaceous in the Western Interior. Here I only repeat overall figures for nonavian dinosaurs, and then briefly review what we know of the bird record. Over the last 10 million years of the Cretaceous in the northern part of the Western Interior, the number of nonavian dinosaurs dropped from 48 to 23, a 52% drop in species diversity. No definite pattern of change for nonavian dinosaurs can be detected for the last few million years of the Cretaceous, but the record is too poor to determine if disappearances were stepwise or abrupt.

The record of Cretaceous birds remains fragmentary; nevertheless, considerable fossil and molecular work has occurred in the past dozen or so years.[10] Two ideas have emerged that appear to have general agreement. First, the crown group of birds, or more formally, Aves, appears to have originated before the K/T boundary. This group is Neornithes, or literally, "new birds." Second, all birds that are stem taxa to Neornithes became extinct before or at the K/T boundary. Beyond these two points there is little agreement, especially between molecularly, anatomically, and paleontologically based results. Chapter 6 deals with similar issues for mammals, in some detail. For birds I limit myself almost exclusively to what happened to them across the K/T boundary.

Starting from one side of the spectrum, ornithologist Alan Feduccia has argued that Neornithes do not originate and radiate until after the K/T boundary. In one of the more inventive phylogenetic diagrams I have seen, Feduccia shows not only the origin and radiation of Neornithes after the K/T boundary but also a very substantial radiation and extinction of Enantiornithes.[11] These are a curious lot; in fact, the name, which was coined by Cyril Walker for this group and probable clade of birds, means "opposite birds."[12] While enanti-

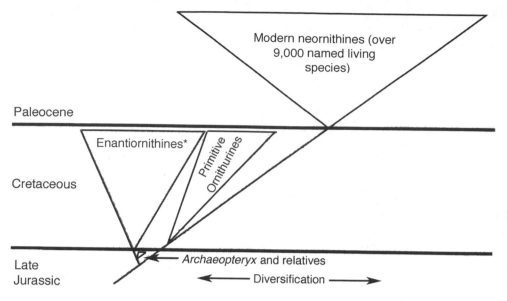

Modern neornithines (over 9,000 named living species)

Paleocene

Enantiornithines*

Primitive Ornithurines

Cretaceous

Late Jurassic

Archaeopteryx and relatives

← Diversification →

*30 named species and only one named from the latest Cretaceous

Fig. 4.3. Simplified version of Feduccia's 1996 diagram portraying enantiornithines as if they are known to be a diverse group in the Cretaceous.

ornithines certainly appear to be ecologically diverse, there are only 30 named species,[13] and to my knowledge, there is only one named species from the latest Cretaceous, and I happen to have found the single specimen upon which it is based.[14] This is a far cry from the radiation in Feduccia's diagram, which makes them a close second to neornithines, which are known from more than 9,000 species. The diagram thus vastly exaggerates what we actually know about the extinction of birds at the K/T boundary (fig. 4.3).

Using the fossil record, Sylvia Hope indicated that three ordinal or higher level clades of neornithines are known from the Cretaceous, while six other clades can be argued to have been present in the Cretaceous as well, based on "guilt by association."[15] This means that if one clade is recognized in the fossil record, its nearest equivalent sister taxon must also have originated. These are appropriately known as ghost lineages, or ghost clades. For neornithines, the major clades that Hope recognized based on the fossil record are Galliformes (turkeys, chickens, and pheasants), Gaviiformes (loons), and Pelecaniformes (frigate birds, pelicans, boobies, and cormorants). The ghost lineages she recognized are Palaeognathae (ostrich, emu, and kiwi), Anseriformes (ducks, geese, and swans), Gruiformes (rails, cranes, and bustards), Charadriiformes (gulls, sandpipers, and plovers), Procellariiformes (albatrosses, shearwaters, and petrels), and Passeriformes (perching birds, comprising 50% of modern bird diversity).

In a different tabulation, Luis Chiappe noted all the above clades except Palaeognathae and Passeriformes, and with the addition of Psittaciformes (parrots), as having been reported from the Late Cretaceous, but he was quick to note that most of these occurrences are based on individual bones that are

hard to identify. Somewhat more complete specimens suggest the presence of anseriforms, and by the ghost lineage implications noted above, palaeognathes and galliforms as well. Echoing what we see for eutherian mammals, Chiappe noted that unquestioned members of modern bird clades are not known in the fossil record until the early Tertiary.[16]

At the other end of the spectrum we have molecular evidence that places the split between two great clades of neornithines, or modern birds—paleognaths (ostriches, emus, kiwis) and neognaths (all other living birds)—as far back as 120 million years ago. This would also mean that most or all lineages of neornithines were around by at least 100 million years ago.[17] Unlike the case for eutherian mammals, for which we have a much better fossil record and thus can successfully challenge the molecular orthodoxy that modern eutherian clades appeared over 100 million years—as I do in Chapter 6—the fossil record is still too fragmentary to address this issue. Further, and despite what some may argue, the record is far too poor to defend any particular pattern of avian extinction at the K/T boundary. As Chiappe noted, "If, in the face of a far richer fossil record, the extinction rate of the last nonavian dinosaurs still lingers unclear, the question of how abruptly the archaic Mesozoic lineages of birds disappeared will likely remain unanswered for years to come."[18]

REPTILIA (TESTUDINES)

Although I have not worked very much with turtles since my PhD research in the mid 1970s, these animals remain my favorite after mammals. By any standard they are some of the best survivors of the K/T boundary. In 1996, of the 17 species (some unnamed) of turtles that I recorded for the Hell Creek Formation, only two did not survive the K/T boundary. Interestingly, only one of these nonsurvivors, *Basilemys*, appeared to be a terrestrial, tortoiselike form. Considerably more work on these turtles has been done since 1996.

Pat Holroyd and her colleagues have taken the lead on this work.[19] Unlike my studies, theirs has examined densely sampled assemblages throughout a 93 m section of the Hell Creek Formation, encompassing something like the last 1.8 million years of the Cretaceous. They can now record a minimum of eight families and 20 genera of turtles, certainly among the taxonomically richest turtle assemblages known. With this better controlled data they were able to establish temporal ranges of turtle species in a local setting with good time and stratigraphic control. Using paleomagnetic data they could tie their sections in eastern Montana with a section in North Dakota where there is good data on megafloras and continental climate change. Except for very rare species, they found that the taxonomic composition was consistent through the vertical extent of the formation. Relative abundances did vary greatly, with chelydrids (snapping turtles) and plastomenine trionychids (soft-shelled turtles) especially common in the lower part of the formation. Higher in the section, no groups were dominant. The greatest numbers of species were found higher in the Hell Creek Formation, correlating to a warming trend 400,000–500,000 years before the K/T boundary.

REPTILIA (SQUAMATA)

In the Hell Creek Formation of eastern Montana there is one rare boidlike snake and 10 species of lizard representing about seven families, depending whose classification is used. The pattern of extinction at the K/T boundary presents an interesting paradox. The fate of the snake is unknown, but all except two of the lizard species disappear at the K/T boundary. In contrast, six of the lizard families (Anguidae, Helodermatidae, Scincidae, Teiidae, Varanidae, and Xenosauridae) are extant, and the seventh (Necrosauridae) is extinct but survived into the Tertiary.

What is one to make of all higher lizard taxa surviving the K/T boundary but the eight of 10 species becoming extinct in the Western Interior? One possibility is that these extinct species are incorrectly assigned to extant families, but there is no evidence to support this view. Among the tetrapods in the Hell Creek fauna, lizards show the greatest diversity in adult body size, ranging from about 15 cm in the smaller teiids to over 3 m for the possible varanid *Palaeosaniwa*. Also, some of these lizard families have a wide range of ecological preferences, so using the lizards to circumscribe aspects of ecology in the latest Cretaceous is not possible. From what is known, most of the lizards were terrestrial, although a semiaquatic habitus cannot be precluded, such as among the teiids, which today include semiaquatic species. As to diet, again the little that is known suggests insectivory through carnivory, depending upon size. There is a hint of some variety, such as the heavy, crushing teeth of in the an-

guid *Odaxosoaurus,* which suggests that it may have fed on heavy-shelled arthropods or mollusks.[20]

Given the apparent diversity of at least size in the Hell Creek lizards, the high levels of extinction at the K/T boundary are somewhat enigmatic because the families survived elsewhere. It certainly argues that whatever eliminated these lizards in the Western Interior was a local or regional event and not a single, sudden, global event.

REPTILIA (CHORISTODERA)

One group of reptiles from the Late Cretaceous that is not familiar to most people is Choristodera, or champsosaurs. They superficially resemble narrow-snouted crocodilians, such as the extant Asian gavials or gharials. They are, however, only distantly related to crocodilians. In fact, we do not know exactly where they fit among other reptiles.[21] About the best we say is that they are diapsids, which really does not say much because among all living reptilian lineages, most are diapsids (including birds and possibly even turtles). The term refers to two openings or, more correctly, to the surrounding bony arches behind the bony eye socket that all diapsids or their ancestors possessed (see fig. 2.1).

Skeletal reconstructions suggest that champsosaurs held their limbs close to the body and swam with strong flexions of the body. Unlike crocodilians, which can bring both eyes and narial openings just above the water to stalk prey, champsosaurs' eyes remained well below water as the tip of the snout broke the surface to breath. This suggests that champsosaurs may have relied on a more strictly aquatic diet.[22]

Even if we do not have a good phylogenetic context for champsosaurs, we have very well-preserved material. Although complete specimens are known but not common, skeletal fragments, especially their distinctive vertebrae, are very common fossils from both the Cretaceous and Paleocene sides of the K/T boundary. From the known material, we can only assess the presence of one species.

ELASMOBRANCHII (SHARKS, SKATES, AND RAYS)

It may at first seem odd that elasmobranchs (sharks and their relatives) have been found in the freshwater deposits of the Hell Creek faunas of eastern Montana. Why they disappeared at the K/T boundary in eastern Montana is discussed more fully in Chapter 5, but suffice it to note here that while elasmobranchs are predominately marine or brackish water creatures, they quite often travel many kilometers inland into fresh water. Still fewer species are restricted to freshwater. There are, however, notable exceptions. Even those elasmobranchs that travel into freshwater must return to salt water to breed, not unlike some bony fish, including some species in the salmon family, that breed in freshwater but live much of their life in salt water.

In 1996 it was reported that there were five elasmobranchs from the Hell Creek Formation in eastern Montana, including one and possibly two sawfishes, a possible guitarfish relative, and two sharks. None was reported from rocks younger than the latest Cretaceous.[23] This still appears to be the case. As discussed in Chapter 5, the geological evidence strongly points to the loss of a marine connection at the end of the Cretaceous, but with a return of this connection in mid Paleocene, when as many as 15 new taxa are known from nearby regions.[24]

ACTINOPTERYGIANS (RAY-FINNED FISHES)

Almost half of the vertebrate species diversity today is composed of actinopterygians, or ray-finned fishes, most of which are marine. Their name comes from the fact that their fins are supported by lepidotrichia, or "fin rays," rather than being fleshy, lobed fins, which characterize the class Sarcopterygii. This should really be the Age of Fishes, not Mammals.

In the Late Cretaceous freshwater deposits in Montana these fish were certainly important, but unfortunately are not thoroughly studied. In 1996, 15 species of actinopterygians were reported from the Late Cretaceous Hell Creek Formation of northeastern Montana, nine of which (60%) survived the K/T boundary.[25]

Interestingly, seven of the 15 fish species—sturgeon, paddlefish, bowfin, and gar—remarkably still have living descendants in the Mississippi River drainage. The remaining eight species are teleosts, overwhelmingly the most diverse group of living fishes. This was not the case for the Hell Creek freshwater fish fauna, for two reasons. First, teleosts had not yet undergone their tremendous evolutionary radiation. As noted, even today in the Mississippi River drainage, a ghost of the Hell Creek fish fauna remains with the presence of paddlefish, sturgeon, gar, and bowfin. Second, the Late Cretaceous record is rather fragmentary compared to that for the Tertiary, and many Late Cretaceous teleostean fossils remain to be examined.

This view was supported by a recent study of isolated teleost elements from the Hell Creek Formation that demonstrated a higher level of diversity of teleosts than usually thought. Further, it appears that there is a still understudied but clear shift in fish faunas between the 75-million-year-old freshwater vertebrate faunas of the Western Interior and the 10-million-years-younger Hell Creek Formation fish fauna.[26]

AMPHIBIA

In 1996, eight species of amphibian—one species of anuran (frog) and seven of caudates (salamanders)—were reported in the Late Cretaceous of the Western Interior, with the same eight species in the Early Paleocene.[27] They had a collective yawn at whatever happened at the K/T boundary, but as we will see, they may have been responding to climatic changes before the boundary.

The majority of living amphibians spend part of their life as an exclusively aquatic larval stage and an adult stage that may or may not take place on land. Their presence is strongly indicative of the presence of fresh water and usually the lack of salt water, which kills most of them. Reproduction must take place in water or in a very wet setting. Phylogenetically, Amphibia is the sister clade to Amniota, which includes Reptilia (including birds) and Mammalia. Amniotes have escaped the watery reproductive phase by producing an aquatic environment for the developing embryo within an egg laid on land or within the body of the female.

Most past studies have concentrated near the top of the Hell Creek Formation and across the K/T boundary into the Paleocene Tullock Formation. A recent study by Carter and Wilson tracked patterns of amphibian paleocommunties through much more of the thickness of the Hell Creek Formation.[28] At the generic level, the study examined changes in taxonomic composition, richness, and relative abundance, during the last two million years of the Cretaceous in northeastern Montana. The results of the study included 53 vertebrate microfossil localities spanning about 93 m of the stratigraphic section. The authors found little change in the taxonomic composition for amphibians during the last two million years of the Cretaceous. Seven genera of caudates (salamanders) persisted through the formation. Relative abundances of some genera fluctuated during the last 500,000 years of the Cretaceous, notably with a high relative abundance of the salamander *Opisthotriton* and low relative abundances of salamanders *Scapherpeton* and *Habrosaurus*. The authors found that amphibian fluctuations correlated with changes in relative abundances of mammal species. Interestingly, the authors pointed out that, whereas there are measurable responses to climate changes by the amphibians during the latest Cretaceous, their response at the K/T boundary was muted, which in turn has implications for the selectivity and causes of extinction.

NONVERTEBRATES
Plants

Ideas on plant extinction at the K/T boundary have changed radically in the past few years. Over the years, the paleobotanist Kirk Johnson has made extensive megafloral (mostly leaf) collections, most notably in the northern part of the Western Interior.[29] In 1992 he reported an almost 80% extinction of the megaflora across the K/T boundary.[30] Fast-forward to 2004, when Johnson and his colleague Peter Wilf presented a quite different portrait of megafloral extinction across the K/T boundary. They found a much lower megafloral extinction—57%—which they regarded as the maximum percentage of plant extinction, down to as low as 30% if pollen are used instead of the megaflora, based on sites from southwestern North Dakota. Pollen offer tighter stratigraphic control in part because they are very small and are often very numerous, but pollen offers less taxonomic resolution than megafloral remains. The authors indicate that "the 57% estimate is significantly lower than previous megafloral observations, but these were based on a larger thickness of latest Cretaceous strata, including most of a globally warm interval, and were less sensitive to turnover before the K/T. The loss of one-third to three-fifths of plant species supports a scenario of sudden ecosystem collapse, presumably caused by the Chicxulub impact."[31] One wonders how low the extinction percentage would need to be to begin to question this extinction scenario. Of course, more important than citing a specific percentage for assessing a cause is determining how rapid the extinction was.

Another paleofloral study of the Western Interior, by Thompson and colleagues, found that the extinction percentage of pollen and spore species across the K/T boundary could be as high as the 30% level found by Wilf and Johnson elsewhere or it could be as low as 15%.[32] While not denying the importance of an extraterrestrial impact at the boundary, Thompson and colleagues argue that environmental changes in the millennia before the impact and the role these changes may have played in the differential extinctions have not been well examined.

These authors found a decrease from 44 to 11 pollen taxa of dicotyledonous angiosperms in the uppermost 3.5 m of Cretaceous sediments.[33] Importantly, there were no changes in the deposition of sediments or the way the pollen was preserved that could account for this decline. Similar declines have been reported in this upper interval at other localities throughout the Western Interior, although species lost at one site may persist elsewhere. This indicates, according to this study, that the declines in species diversity are local events, not a general gradual extinction.

This biostratigraphic/geographic pattern indicated that the latest Cretaceous plant communities were becoming more heterogeneous, which in turn altered the habitats for some vertebrates that depended on these plants. Local disappearances of the plant species rendered these particular vertebrates more vulnerable to extinction from major environmental disturbances, such as an extraterrestrial impact. As the authors note, these more complex environmental interactions better explain the differential vertebrate extinctions at the K/T boundary. I refer to this study again in Chapter 5, as the patterns of decline and extinction for plants discussed here are equally relevant to demarcating likely causes of extinction.

The final paleofloral site in western North America to discuss, the Castle Rock site, which has been described as "a hyperdiverse, 64-million-year-old tropical rainforest megaflora from the western margin of the Denver Basin in Colorado, suggested that the Paleocene megafloral record, however extensive, was still incomplete. Compositionally similar but an order of magnitude more diverse than other Paleocene floras, the Castles Rock site presented a conundrum that remains unresolved today."[34] Later the authors of this quote, Doug Nichols and Kirk Johnson, note that the Early Paleocene Castle Rock rainforest was diverse, but not as diverse as Cretaceous floras in the Western Interior. They note that implications for the K/T boundary are unclear.[35] I would agree

with this uncertainty if one assumes, as these authors do, that the K/T extinctions were equally cataclysmic on a global scale. The meaning of the Castle Rock flora comes into focus with the realization that the causes and patterns of extinctions were multiple, not single. The plants and animals were responding to a complex set of changes.

The preceding discussion concerns paleofloras from the Western Interior of North America, but do we have relevant paleofloras from elsewhere in the world that shed light on the patterns of extinction and survival across the K/T boundary? One such megaflora from southern Argentina has been described by, among others, two of the authors of one of the Western Interior studies noted above, Wilf and Johnson.[36] Recall that a megaflora includes larger specimens, such as leaves, fruits, flowers, seeds, and cones. This particular megaflora included 36 species of angiosperm, or flowering plants, ferns, and conifers. The Argentinean megaflora in question does not include the K/T boundary but, at some 61.7 millions years ago, is about four million years into the Paleocene. This paleoflora, which represents low-lying floodplains in a humid, warm-temperate climate, is paleoenvironmentally quite comparable to Paleocene floras from the Western Interior of North America, yet this South American flora is more than 50% richer in species than floras in the Western Interior. The authors suggest that this indicates a more vibrant terrestrial ecosystem in South America than in the comparable North American floodplain environments just four million years after the K/T boundary. It further indicates high floral diversity in South America 10 million years earlier than previously thought.

The authors speculate that the considerable disparity in taxonomic richness at this time in the Paleocene between comparable North and South American floras could involve reduced effects of an extraterrestrial impact in this part of South America because of greater distance from the Chicxulub impact site, faster recovery or immigration rates after the K/T boundary, or initially higher latest Cretaceous diversity. Whichever of these ideas is correct, and they could all be correct, these issues once more point out the differential effects of extinction and survival at the K/T boundary and the much greater likelihood of multiple causes of extinction. As noted for mammals earlier, this differential pattern of extinction and survival also holds true in that eutherians fare much better than metatherians in survival at the K/T boundary in the Western Interior, but when we first see them in the earlier Paleocene in South America, both have done well, with eutherians mostly herbivores and metatherians omnivores and carnivores.

Invertebrates

The term *invertebrates* is one of those unfortunate catchall words meaning anything that is not a vertebrate. I tell my students that using it makes about as much biological sense as referring to all people in the world as Californians and non-Californians makes ethnographic sense. With the notable exception of freshwater snails and bivalves, we have a rather paltry record of invertebrates in the Western Interior. Joe Hartman has studied these freshwater bivalves, also known as unionoid bivalves, for many years. To most of us uneducated in these creatures, a typical mussel or clam would be the best, if slightly inaccurate, image.

In his most recent study with a colleague, Henning Scholz, Hartman examined the patterns, causes, and ecological significance of extinction of these bivalves near the Cretaceous/Tertiary boundary of the Western Interior.[37] They note that the Hell Creek Formation taxonomic diversity is exceptionally high, with 30 some species of bivalves, not unlike the diversity of the modern Mississippi River drainage. This fauna undergoes a significant turnover near the Cretaceous/Tertiary (K/T) boundary, and the decrease in taxonomic diversity in the interval spanning the K/T boundary is associated with a significant shift in morphospace occupation, that is, different forms came to dominate. Noting that this is congruent with a decrease in habitat stability, they discuss four processes and events as possible causes of this decrease: the extraterrestrial impact at Chicxulub, global climate changes

in some way related to Deccan volcanism, changes in the emerging Rocky Mountains, and global changes in sea level. They concluded that the extraterrestrial impact was not the major event for bivalve extinction, arguing that effects of the emerging Rocky Mountains and global changes in the sea level were much more important in explaining bivalve species turnover at the end of the Cretaceous.

While I am in agreement with these authors on much of their interpretation, there is one point regarding the geological setting with which I may disagree. They think that an increase of sea level, which probably began in the Late Cretaceous, further away from the Western Interior had reached their study area, mostly in western North Dakota, by the Early Paleocene (the Cannonball Sea). As is discussed more extensively in Chapter 5, both sharks and relatives, as well as geology, suggest that the shallow seas had left the Western Interior by the K/T boundary, not to return until the mid-Paleocene, when both sharks and relatives and marine sediments also returned. I do not know if these differences in interpretation of the fossil and rock record would change the authors' conclusions in any substantial manner regarding the fates of bivalves.

5 Causes of Extinction at the K/T Boundary

THE SUBTITLE OF THIS VOLUME, "How the Fall of Dinosaurs Led to the Rise of Mammals," might suggest to some readers that I would implicate mammals in the extinction of nonavian dinosaurs.[1] This is a very unlikely possibility. Nevertheless, in the greater scheme of evolutionary history and on a shorter ecological scale, we know or can surmise that sometimes very distantly related animals of very different sizes can and do compete for resources. We know, for example, that native birds and reptiles are greatly imperiled when rats or other mammals escape or are intentionally released on oceanic islands; devastation and extinction is almost inevitable. Similarly, when plagues of locusts, or more correctly, grasshoppers descend in swarms of millions they lay the earth bare, decimating the food for man and beast in the surrounding area.

The realization that quite different animals of quite different sizes can and do influence one another in the web of life has been known a long time. For instance, one of my favorite quotes about the interactions of species at different scales is that of Charles Darwin in *On the Origin of Species* in which he waxes whimsically about the interaction of flowers, humble bees, mice, and cats:

> I am tempted to give one more instance showing how plants and animals, most remote in the scale of nature, are bound together by a web of complex relations. . . .
> The number of humble-bees in any district depends in a great degree on the number of field-mice, which destroy their combs and nests. . . . Now the number of mice is largely dependent, as everyone knows, on the number of cats; and Mr. Newman says, "Near villages and small towns I have found the nests of humble-bees more numerous than elsewhere, which I attribute to the number of cats that destroy the mice" Hence it is quite credible that the presence of a feline animal in large numbers in a district might determine, through the intervention first of mice and then of bees, the frequency of certain flowers in that district![2]

Just as in Darwin's example, there is certainly evidence from modern biotas indicating that one group of animals has caused decline or local extirpation of another group animals, but demonstrating causality for total extinction is harder to come by, especially for extinct species. (Human-caused extinctions are a notable exception.) A very good case has been made that early rodentlike primates and true rodents may have outcompeted a previously successful group, the multituberculates, in the early Cenozoic over millions of years.[3] Another somewhat obscure but quite probable case of competition leading to extinction is that of the early Tertiary apatemyids known from Europe and North America. They had chisel-like incisors, probably for gnawing away bark, and several elongate digits, probably used for probing for insects. Their habits can be surmised because the living but endangered aye-aye, a Malagasy relative of lemurs, and the striped possum, an Australian marsupial, both have gnawing incisors and elongate digits for removing insect prey from within trees.[4] The nearly global spread of woodpeckers in the Tertiary may have spelled doom for the apatemyids. Possibly the aye-aye and striped possum survived because there are no woodpeckers on Madagascar or Australia.

The main charges leveled against mammals as possible exterminators of nonavian dinosaurs are that they could have eaten the eggs of these dinosaurs or that they outcompeted them ecologically. There is really no question, based on today's biota, that there may have been some interactions between mammals and nonavian dinosaurs, but these were in all probability limited because most dinosaurs (here including birds) were likely diurnal (active in the day), whereas most mammals, as they are today, were mostly nocturnal (or active at night). This may come as a surprise to some, but the vast majority of mammals are today, as in the past, small, nocturnal, solitary creatures.[5] Further, why and how could mammals devastate nonavian dinosaur numbers after more than 100 million years of coexistence by eating their eggs or competing for food? The only obvious larger scale ecological comparisons between nonavian dinosaurs and mammals are that nonavian dinosaurs occupied the large, terrestrial vertebrate niches occupied by mammals today. Only after nonavian dinosaur extinction could mammals radiate into the ecologically staggering array of forms we see today. As we shall see in last chapter, mammals unquestionably benefited from the extinction of nonavian dinosaurs, but did they cause it? I do not think so. Not to despair, however, as there are many more extinction hypotheses out there.

If mammals can almost certainly be exonerated in the extinction of nonavian dinosaurs, what are the likely causes or culprits? The British paleontologist Alan Charig had tabulated some 80 hypotheses of nonavian dinosaur extinction.[6] I think Charig was somewhat free in his estimates as to the number of truly different hypotheses. Nevertheless, they are many, no matter how one counts them. Over the years I have not attempted to keep track, but I do not recall any truly new hypotheses emerging, except, possibly, one proposed by Doug Robertson and his colleagues as a corollary of an extraterrestrial impact—more on this later. To this day the major testable components of hypotheses of the K/T extinctions as ultimate causes are volcanism, marine regression, and extraterrestrial impact.[7]

There are other hypotheses ranging from the ridiculous (alien overhunting, constipation, drowning in Noah's flood)[8] to the interesting but untestable, such as the spread of some infectious disease. This is not the case with hypotheses resting on some aspects of volcanism, marine regression, and an impact near or at the K/T boundary. There is overwhelming geologic and correlative paleontologic evidence for three events: for a span of about eight to nine million years sur-

rounding the K/T boundary the massive Deccan Traps were forming on the Indian subcontinent; during the waning hundreds of thousands of years of the Cretaceous the shallow seas that had covered vast reaches of the continents were draining away; and at, and probably defining, the K/T boundary, a massive projectile from space smashed into what is today southern Mexico.

There is quibbling over various aspects of these events. For example, it not clear when the eruption of the Deccan Traps was most severe, but possibly round 66 million years ago, spanning the K/T boundary. Similarly, there is some evidence that the seas lingered longer on the North American continent than was once thought and that the seas may have begun to return before the K/T boundary. As to the impact, it is not clear what the object was that struck earth, how large it was, and if there may have been several impacts leading up to the K/T boundary.[9] This said, the evidence that these three physical events occurred is overwhelming, much more in the realm of observations than hypotheses. I will discuss some of the caveats later. The real issue and greatest area of controversy is whether just one of these events or some combination of all three lead to the K/T boundary extinctions. This has become known as the single-cause versus multiple-cause explanation for the K/T extinctions.

In 2004 David Weishampel and his colleagues published the massive second edition of their encyclopedic work *The Dinosauria*. David Fastovsky and I coauthored the chapter on nonavian dinosaur extinction. He argued his single hypothesis for nonavian dinosaur extinction, while I argued the multicause hypothesis. I cannot speak for him, but I found it a good learning experience, and even though the chapter is slightly out of date, I think it still offers a succinct presentation of these competing hypotheses. One of the things we attempted to do, and I believe succeeded in doing, was to indicate when we agreed concerning the geologic and paleontologic record and when we disagreed, sometimes based on the same observations. In Chapter 1 of this volume, using extensive data from *The Dinosauria*, I detailed some of the rather disparate views on the fossil record of nonavian dinosaurs. I mention these as relevant here, although this chapter focuses on hypotheses and possible consequences.

EXTRATERRESTRIAL IMPACT: A SINGLE ULTIMATE-CAUSE HYPOTHESIS OF EXTINCTION AND ITS PROXIMATE COROLLARIES

There is little doubt within the geologic community that some object from space slammed into the earth about 66 million years ago at and in effect defined the K/T boundary.[10] When Luis Alvarez, his son Walter, and their other colleagues proposed in 1980 that an asteroid impact was the cause of the mass extinctions at the end of the Cretaceous, there was much consternation and doubt that such an event had occurred, let alone caused extinctions. As a recently minted PhD from UC Berkeley who had worked on mammals across the K/T boundary in the Hell Creek region of eastern Montana, I was among the naysayers, primarily because I did not see the need for such an impact to explain the extinctions. I do not know when I became convinced that an impact had occurred, but my best guess is when the hallmark of the boundary, the platinum group metal iridium, began to show up not just in marine sections in Italy, Denmark, Spain, and New Zealand but also in terrestrial sections in Montana, notably at the top of the famous Hell Creek Formation at the K/T boundary. Another piece of evidence was the discovery of so-called shocked quartz, which shows multiple planes of fracture that leave their mark in the quartz mineral stishovite formed in this process. While very high-pressure events on Earth might form it, the real culprit appears to be instantaneous, high-pressure events, such as extraterrestrial impacts. In fact, it was first named from occurrences at Meteor Crater, Arizona.[11]

Thus, I was becoming convinced of an impact even before the discovery of the Chicxulub impact crater in the Yucatan was formally proposed in 1991.[12] The issue remained, why was such an impact necessary for these mass extinctions? Couldn't terrestrial events explain them equally well? The way for a paleontologist such as myself to test the various extinction scenarios was to track the change of vertebrate species across the K/T boundary and ask how the patterns of extinction and survival matched with the various extinction scenarios. I began this work with my colleague Laurie Bryant in the late 1980s and have continued these tabulations off and on over the years. The results, as I discussed in the last Chapter 4, show differential levels of extinction across different vertebrate groups, leading me, as well as others, to the conclusion that one single cause, no matter how dramatic, could not explain the disparate pattern of extinctions at the K/T boundary. Others do not agree.

In our 2004 chapter for *The Dinosauria*, my coauthor, David Fastovsky, wrote, "The single-cause argument for the K/T extinctions is fundamentally a parsimony argument. . . . Any hypothesis purporting to explain events at the K/T boundary must meet two criteria: (1) the hypothesis must be testable, and (2) the hypothesis must be able to explain as much of what is known about the boundary as is possible." He went on, "In this sense, the argument that a variety of causes (asteroid impacts, volcanism, regression) produced a variety of effects on a variety of different organisms is unsatisfying. If more 'causes' were known would we then be able to explain better the effects? The multiple-causes-multiple-effects viewpoint is a default: simply because these events are known to have occurred is not *a priori* reason to consider them causes. Many of these events occurred many times pervasively—even in conjunction—and didn't cause a mass extinction."[13]

In effect, those who argue for multiple causes at the K/T boundary are accused of the crime of confusing correlation and causation. The issue is that "correlation does not imply causation," meaning that because two variables are correlated does not mean that one causes the other. David would be correct if proponents were trying to tie all events at the K/T boundary with the mass extinctions. They are not; rather, they are showing that volcanism, marine regression, and extraterrestrial impact are necessary, but not separately sufficient, to explain the differential pattern of extinctions at the K/T boundary; together they come much closer to explaining these extinction patterns. The frequent pronouncement that because there was an impact at the K/T boundary, it must have been the sole cause of the mass extinctions, is false, as I will explain.

I am certainly not alone in arguing that there are multiple causes of extinction not only at the K/T boundary but also elsewhere throughout the geologic past. In 2008 Nan Arens and Ian West examined the idea that extensive volcanism, in combination with an extraterrestrial impact, might be mechanisms of mass extinction. They tested this idea by examining generic extinction at 73 levels in the Mesozoic and Cenozoic eras. They found that elevated extinctions were more frequent when there was both volcanism and an extraterrestrial impact, neither of which caused elevated extinctions by itself. Even with this, they found that the combination of volcanism and impact alone could not explain all intervals of elevated extinction. They discussed other factors, such as climate and sea-level change, that alter community composition by placing longer-term stress on ecosystems. As I discuss later in this chapter, it is this longer-term stress in the terrestrial realm that put vertebrates and other species at risk. I quote from Arens and West's conclusion: "Our analysis suggests that single causes for mass extinction—although appealing in their simplicity—may be inadequate to explain the detailed data now available for many episodes of mass extinction."[14] They refer to this as their "press-pulse model," in which the longer-term events are the "press" part of their model and sudden events are the "pulse."

Single causes for the K/T extinctions are demonstrably inadequate; nevertheless, this has not stopped their promulgation. A 2010 article was touted as the definitive consensus review supporting the Chicxulub asteroid impact as the single cause for the terminal Cretaceous extinctions.[15] This impression was bolstered because the article was published in one of most widely read scientific journals, *Science*, and it boasted a roster of 41 authors. I was not alone in having colleagues and friends ask whether this was the definitive review article for the K/T extinctions. With considerable hubris this was acclaimed the "dream team,"[16] yet analysis of the various organisms affected at the K/T boundary was superficial. Especially telling, dinosaurs—the poster children of the K/T extinctions—were invoked three times in the review, yet not one of the 41 authors had worked on dinosaurs. Even more shocking, there was not a single vertebrate paleontologist among the authors; only paleobotanists represented the entirety of work in the terrestrial realm. Possibly *Science* was unaware of the extremely skewed composition of the authorship of the review, but then again, they may have known full well. Whatever the truth may be, the response to the review from the paleontological and geological community was swift and quite negative.

Within 48 hours of the review's publication, 29 of us drafted a letter of rebuttal that, to their credit, *Science* agreed to publish along with responses of others.[17] Unlike the authorship of the review, which included those working mostly on marine microfossils, geophysics, stratigraphy, and plants, our letter included researchers working on mammals, dinosaurs, birds, pterosaurs, crocodilians, turtles, lizards, sharks, plants, freshwater and marine invertebrates, sedimentology, sequence stratigraphy, volcanology, and statistical methods at and across the K/T boundary. What this broadly representative group of researchers concluded was that the pattern and timing of extinctions at the K/T boundary overwhelmingly pointed to multiple causes.

The Supposed Significance of Impactor Size

In the original paper in 1980, the Alvarez group argued that, taken together, the concentrations of iridium, data on craters that earth-crossing asteroids produce on the earth, the 1 cm boundary layer composed of material that fell from the stratosphere, and scaled-up estimates of opaqueness produced by 1883 Krakatoa explosion all indicated that the size of the asteroid impact was on the order of 10 km in diameter. When found, the ~180 km diameter Chicxulub crater corresponded to estimates of craters that would be produced by an asteroid of 10 km in diameter.[18] The estimate for the crater is now slightly smaller, at 170 km,[19] but nonetheless still very large.

If the single-cause hypothesis of K/T extinctions is correct, then other large impacts should show extinction levels commensurate with their sizes according to proponents of impact-induced extinction. In the early 1990s, David Raup calculated a "kill curve" that estimated ~65% species extinction for an impact crater of ~140 km diameter and ~30% extinction for one of ~60 km.[20] More recently it has been shown that Raup's original "'kill curve' must, at minimum, be recalibrated to exclude extinction effects from craters smaller than ~100 km in diameter."[21] Why would this be the case? The answer is that smaller craters, of 100 km and less, some of which Raup thought were related to extinctions,[22] have been found to have occurred at times other than those of major extinction. This calls into question the basic premise of calculating extinction rates based on estimated crater sizes.

For example, the two largest craters known for the Cenozoic are the Popigai crater, of 100 km diameter at 35.7 million years ago, and the Chesapeake Bay crater, of 90 km at 35.5 million years ago.[23] Another likely crater of the same age and possibly related to the Chesapeake Bay crater is the Toms Canyon crater, at 20–22 km.[24] Interestingly, the Popigai and Chesapeake Bay craters are "the two largest Cenozoic impact craters originated within a time span of just a few thousand to hundred thousand years."[25] For the Chesapeake Bay and Toms Canyon craters, Poag and his colleagues found "no immediate major extinction events." They did note that "there is evidence, however, that the Chesapeake Bay and Toms Canyon impacts helped initiate a long-term pulse of warm global climate, whose eventual dissipation coincided with an early Oligocene mass extinction event, 2 Ma *after* [the authors' italics] the impacts."[26] They further note that "in conjunction with the essentially isochronous Popigai impact in northern Siberia" there was a Late Eocene warming trend.[27]

While the warming trend in the Late Eocene may be related to these multiple impacts, it may not, and it is interesting that the Chicxulub impact is often implicated in a sudden, prolonged freeze, not a very long-term warming trend. Does this mean that cratering at or below 100 km causes warming, but well above 100 km causes a cold snap? I must make it very clear that I do agree that major impact events almost certainly cause major perturbations to the globe, but we are not at the point, as some would have us believe, that we know with great certitude that the Chicxulub was the single, catastrophic cause of the K/T mass extinctions. We can even broaden our conclusions regarding large impacts, namely, that none of the other four mass extinctions that have occurred during the last 545 million years of the Phanerozoic Eon have been linked convincingly to an extraterrestrial impact. As we will see, however, marine regressions (the loss of shallow seas) correlate to all five mass extinctions in the Phanerozoic Eon.

Downsizing the Impactor

There is a final troubling aspect to the claims made for the Chicxulub impact as the single cause of the K/T mass extinctions. Much of the energy extrapolated for these impacts comes from estimates not only of crater size but also of the projectile size. As noted earlier, the most common estimate for the diameter of the Chicxulub crater is now given as 170 km, with estimates of between 10 and 19 km for the size of the projectile. A recent study by Paquay and others examining ratios of isotopes in seawater, notably of osmium and also including iridium data, indicates that the sizes of the impact projectiles have been overestimated.[28] Impact simulations found 15 to 19 km for the K/T Chicxulub crater, 8 km for the Late Eocene Popigai crater, and 3 km for the Late Eocene Chesapeake Bay crater. In contrast, Paquay and colleagues estimated 4.1 to 6 km for the K/T Chicxulub crater and 2.8 to 4.0 km for the Late Eocene Popigai crater. If corroborated, these findings mean that the size of the projectile that struck at Chicxulub has been overestimated by a factor of two to three. This has major implications for determining the amount of energy that was released by this impact and, of course, for the catastrophic global extinctions that are claimed by the single-cause hypothesis of K/T extinctions. Not surprisingly, there have been objections to the

Paquay et al. study, most importantly for consideration here, that their methodology underestimated the size of the projectile.[29] Time and more studies should narrow the window on the size of the projectile, but suffice it to say that there is not incontrovertible evidence for the size and biological effects of impacts at the K/T boundary or anywhere else in the geologic record.

The Angle of Attack

It appears that both the size and the angle of an impact may determine the amount of biological devastation. It is difficult to assess the size of the projectile, and it is even harder to gauge the angle at which it struck earth. After a talk in 2007 in the Department of Geology at my university, I asked the speaker, Christian Koeberl, one of the world's experts on impacts, about the issue of determining the angle of impact at Chicxulub.[30] He responded that in most instances, including that at Chicxulub, the angle of impact cannot be determined with any certainty. Nevertheless, there are widely varying assessments of angle of impact in the literature.

A paper in 1996 by Peter Schultz and Steven D'Hondt argued that asymmetric geophysical signatures suggested a low angle trajectory, of between 20° and 30°, for the Chicxulub impact. They further determined that the direction of travel was from the southeast to the northwest. The biological implications were intriguing. They argued that with this low angle and direction, "biotic extinctions may have been most severe and catastrophic in the Northern Hemisphere. Geographic variation in the magnitude of the Cretaceous-Tertiary (K-T) 'fern spike' and palynofloral extinctions are consistent with the proposed trajectory."[31] Their proposed trajectory would create what they termed "a corridor of incineration" along its path, from southeast to northwest, with radiation well above levels needed to cause spontaneous ignition of plant life. They suggested that the large heat capacity of water could have provided refuge for some plant and animal species. The authors argued the records of pollen and spores were consistent with their hypothesis of a low-angle strike, notably, that available data suggest that palynofloral extinctions in North America were greater than elsewhere, for example, in South America.[32]

Ten years later, Joanna Morgan and her colleagues came to quite different conclusions. They examined the distribution of shocked quartz around the Chicxulub impact site and found that the total number, maximum size, and average size of shocked quartz grains all decreased gradually going away from Chicxulub. Among other things, they found that the size distribution of grains was relatively symmetric outward from the crater. They suggested a high-impact angle of about 45°, possibly to the southeast. They argued that such an oblique impact was likely "far more catastrophic than a sub-vertical one, because greater volumes of volatiles would have been released into the atmosphere," and also "that some of the more extreme predictions of the environmental con-sequences of a low-angle impact at Chicxulub are probably not applicable."[33]

In order to explain the differences with the low-angle strike found by Schultz and D'Hondt, Morgan and her colleagues noted that they had not found as much shocked quartz in the North Pacific as reported by Schultz and D'Hondt, one of these authors' lines of evidence for a low-angle strike toward the northwest at Chicxulub. Morgan and her colleagues also pointed out that in the original 1980 paper, Alvarez and his colleagues had reported "that the ejection angle and earth's rotation produced the high abundance of shocked quartz in the Pacific." Morgan and her colleagues concluded "that comparisons of results from unrelated studies are problematic," leaving the reader pondering what to believe about shocked quartz in the Pacific and its importance in determining the angle at which Chicxulub was struck.[34] Their final conclusion is also puzzling: "Previous studies have predicted that Chicxulub produced a dramatic effect on the earth's climate because it was formed by a low-angle oblique impact into volatile-rich sediments. Our results suggest that these more extreme predictions are unlikely."[35] They do not say which extreme predictions are unlikely, but the clear implication is that the climatic and hence biotic effects have been overplayed for the Chicxulub impact.

Darkening Skies

In their historic 1980 paper, Alvarez and colleagues postulated that when an asteroid struck Earth at the K/T boundary pulverized rock some 60 times the mass of the impacting projectile was injected into the atmosphere. Some fraction remained distributed globally in the stratosphere for several years. They concluded that this "temporary absence of sunlight would effectively shut off photosynthesis and thus attack food chains at their origins." They further noted that on land a "food chain is based on land plants. Among these plants, existing individuals would die, or at least stop producing new growth, during an interval of darkness, but after light returned they would regenerate from seeds, spores, and existing root systems. However, the large herbivorous and carnivorous animals that were directly or indirectly dependent on this vegetation would become extinct."[36]

While this scenario has remained one of the favorites of impact theorists, there have been detractors. In 2002 the geologist Kevin Pope argued that the impact of a 10 km diameter asteroid as hypothesized by Alvarez and colleagues simply was not great enough to inject the amount of dust into the atmosphere to cause what has been called by others a "cosmic winter," an "impact winter," or a "nuclear winter." The last name refers to the supposed similar effects of a global nuclear conflagration. Pope calculated that at least twice the magnitude of fine dust would have needed to be generated in order to suppress photosynthesis, followed by the hypothesized extinctions of vertebrates.[37] Instead, he proposed that the culprit was sulfate aerosols (a contributor to acid rain) produced from the rock hit by the asteroid and ash produced in global fires, both of which caused cooling, which interfered with photosynthesis.[38] As we will see, however, the idea of acid rain, global fires, and global cooling are even less well supported than global darkness caused by dust.

In 1980, Alvarez and his colleagues concluded, "In a general way the effects to be expected from such an event are what one sees in the paleontological record of the extinction."[39] Even if they are correct and Pope is wrong about the amount of dust in the atmosphere, their scenario more poorly explains the patterns of survival and extinction for nonmarine vertebrates other than nonavian dinosaurs. Other causes were required. Some of these other causes have been suggested as proximate causes or corollaries of an impact at the K/T boundary. Among these, some have fared better than others in explaining the pattern of vertebrate survival and extinction across the K/T boundary.

Acid Rain

Many people have heard of acid rain, even if they do not fully comprehend its impact on the environment. It is known to be responsible for the killing of vast stretches of forests in the Northern Hemisphere and the deadening of smaller bodies of water. Although volcanoes can produce the aerosols that result in acid rain, most of the acid rain today is caused by industrialized human activities. In the immortal words of Pogo from the first Earth Day, in 1970, "We Have Met the Enemy and He Is Us."[40]

Recall that how acidic or basic a substance is, is measured by means of a pH scale,[41] normally calibrated from zero to 14, with seven being neutral. Below seven a substance is acidic and over seven it is basic. Each whole value indicates a pH 10 times lower or higher than its preceding value. Each whole pH value below seven is ten times more acidic than the next higher value. Thus a pH of two is 10 times more acidic than a pH of three and 100 times more acidic than a pH of four. When in equilibrium with the atmosphere, water has a pH of 5.6, while pure water (no dissolved gases) has a neutral pH of 7.0. A pH of below 5.0 is considered quite acidic. Rain as low as 2.4 has been recorded, but annual averages in areas affected by acid rain range from 3.8 to 4.4. Acid fogs and clouds from 2.1 to 2.2 have been recorded in southern California and have been known to bathe spruce-fir forests in North Carolina.[42] Aquatic vertebrates (fish, amphibians, and reptiles) are especially vulnerable to acidic waters. The first to suffer from a low pH are those reproducing in water, but if the pH reaches lower than three, adults usually die.

Given these known dire consequences for a low pH on aquatic vertebrates, the hypothesized pHs for acid rains at the K/T boundary are truly astounding. Sulfuric and nitric acid are cited as the usual culprits, the former resulting when large amounts of sulfur dioxide are vaporized from rock at the impact site and the latter when tremendous energy released by the impact allows the combination of atmospheric nitrogen and oxygen. These events are supposed to have driven global

pH levels to below three and possibly between 0 and 1.5.[43] Recall from the chapter dealing with patterns of vertebrate extinctions at the K/T boundary that aquatic vertebrates are some of the best survivors—so much for a global pattern of very low acid rain at the K/T boundary. I also point out that "just so stories"—the acid rains were somehow buffered by neutralizing carbonate rocks, much as a basic solution might provide—are in effect nonstarters.[44] If there was acid rain, its consequences were now much more muted.[45]

A Sudden Chill

In 1991, the paleobotanist Jack Wolfe had a paper in the journal *Nature* titled "Palaeobotanical evidence for a June 'impact winter' at the Cretaceous/Tertiary boundary." I was immediately reminded of Archbishop Ussher's 1650 proclamation on the first page of his *Annales veteris testamenti* (*Annals of the Old Testament*) that the Earth was created on October 23, 4004, BCE. While the date of October 23 is of course beyond belief, or rather testability, Wolfe's June date is not as wacky as it might first sound. Basically, he argued that the reproductive stages of the fossil aquatic plants he recovered suggested death and burial in approximately early June. The real problem is whether there was a brief cold snap or "impact winter" of one to two months, as he argued, only for terrestrial environments but not for marine environments. He further argued that the sediment in and around the plant fossils indicated not one but two impacts—a large distant impact and a closer smaller impact. Other paleobotanists were not too kind in their assessment, most importantly pointing out that Wolfe's pollen identifications were incorrect,[46] which would seem to undermine evidence for a June "impact winter."

Is the idea of an "impact winter" still supported by other evidence? It appears that the idea of such a sudden chill grew out of hypothesizing as to what would happen if and when dust injected into the atmosphere over much of the globe reduced or cut off sunlight, as I discussed earlier. Citing work of atmospheric scientists, two of the original asteroid impact scientists, Walter Alvarez and Frank Asaro, wrote, "The darkness would also produce extremely cold temperatures, a condition termed impact winter."[47] Is there any other fossil data that supports such a cold snap? For vertebrates, the answer is no. If this had been the case, the first vertebrates to be affected would have been the tropical to subtropical ectotherms,[48] such as crocodilians, that cannot tolerate freezing; they become sluggish or immobile at 10°C–15°C. As my friend and colleague Howard Hutchison noted, the northern distribution of extant turtles and crocodilians is limited by temperature.[49] Totally aquatic vertebrates such as fish would be less affected because of the buffering effects of water. We must remember that the northern Western Interior, very much unlike today, was a subtropical to tropical region buffered by the nearby shallow seaway, and thus the extended, subfreezing temperatures advocated by proponents of an "impact winter" would have been devastating to ectotherms living there. Most ectotherms, however, except for sharks and relatives, pass almost unscathed through the K/T boundary.

Further evidence comes from farther north. Bill Clemens and Gayle Nelms described a latest (but not terminal) Cretaceous vertebrate fauna from along the Colville River in Alaska at about latitude 70°N, which was probably at latitude 85°N during the Late Cretaceous, but in either case, well above the Arctic Circle.[50] Thus, any organisms there during the northern winter (which included the plants that were certainly not going anywhere) faced more than three months of dusk or darkness. The authors found that of the 49 genera of amphibians and reptiles that were recognized from the Western Interior, excluding nonavian dinosaurs, none was in their Alaskan fauna. The simplest and likely correct explanation provided by these authors was that endothermic tetrapods (nonavian dinosaurs and mammals) and ectothermic fishes (buffered by water) could deal with the lower temperatures in latest Cretaceous Alaska, while ectothermic tetrapods could not. If the temperature range of 2°C–8°C estimated by Clemens and Nelms was sufficient to exclude ectothermic tetrapods during the latest Cretaceous in Alaska, a severe temperature drop to below subfreezing temperatures at the K/T boundary farther south, in the relatively balmier Western Interior, would have devastated the rich ectothermic tetrapod faunas. This did not occur, casting the idea of an "impact winter" quite a substantial blow.[51]

Tsunamis No More

Tsunamis are an interesting corollary claimed to have been the result of an extraterrestrial impact at the K/T boundary. Tsunamis only became very obvious to most people in 2004, when an earthquake west of Sumatra triggered a devastating wave train (some waves over 30 m high), or tsunami, along the coastlines of the Indian Ocean, killing almost a quarter of a million people. In 1988, tsunami deposits were reported in the well-studied K/T boundary Brazos River sections in central Texas. Other tsunami deposits were later reported in the southeastern United States and around the Gulf of Mexico. Large rocks and other indications of sudden deposition characterize such deposits. When the sections at Brazos River were more carefully studied the putative tsunami deposits were found to be a series storm deposits separated by at least six surfaces in which marine creatures had burrowed. Thus, the unit was not one big tsunami deposit but rather represented a series of storms followed by longer intervals, of as much as 10,000 years, when marine creatures burrowed into the underlying storm deposits. The K/T boundary was found above these storm units, which thus had nothing to do with an extraterrestrial impact.[52] There may have been tsunamis resulting from the K/T boundary impact, but such deposits remain elusive.

Infrared Radiation and Global Wildfire

I save the effects of infrared or thermal radiation for the end, because, as with the scenario of prolonged darkness, it best fitted what we knew of vertebrate extinction and survival at the K/T boundary. Newer modeling, however, casts serious doubt on this hypothesis. This hypothesis, published in 2004 by Douglas Robertson and colleagues,[53] offered one of the more biologically well-thought-out hypotheses of how the impact of an extraterrestrial object could cause the differential pattern of extinctions of nonmarine species at the K/T boundary. This was almost certainly in part because three of the authors were vertebrate paleontologists.

The hypothesis argues that, for a matter of hours following the impact at Chicxulub, Earth was bathed in intense infrared radiation, which resulted from ejecta reentering the atmosphere. The heat pulse would also have directly killed any organisms that were not sheltered. Only organisms that could find adequate shelter in natural cavities, could burrow, or were protected by even a few millimeters of water might have had a chance of survival.

As Norm MacLeod and I have pointed out elsewhere,[54] this hypothesis at first appears to explain why larger nonavian dinosaurian species did more poorly than did some mammals: they could not hide. Large size, however, may not be the major issue. Nonavian dinosaurs were small when hatched, and some may even have been small enough after leaving parental care to survive in some sort of cavity. But in such cases, there would have been enough survivors to form a viable postimpact population to ensure species survival.

Equally problematic for Robertson and his colleagues' scenario, in 2007 a Cretaceous burrowing dinosaur showing both traces of the burrow and fossil remains was recovered form western Montana. As the authors noted, in reference to the Robertson scenario, "The ability among dinosaurs to find or make shelter may contradict some scenarios of the Cretaceous-Paleogene impact event."[55] The authors rightly equivocated because their dinosaur and burrow were found in sediments of early Late Cretaceous age rather than near the K/T boundary. This, however, does show that at least a few nonavian dinosaurs could have escaped the ravages postulated by Robertson and his colleagues, and yet all nonavian dinosaurs became extinct. If this were not enough, in 2009 yet more burrows attributable to nonavian dinosaurs were discovered in southeastern Australia. Again, these were from rocks not of latest Cretaceous age but of the Early Cretaceous.[56]

Almost certainly mammals, which were all small relative to at least adult nonavian dinosaurs, had much larger population sizes. Their numbers in the fossil record indicate this. With a larger population size, coupled with small physical size, more individuals may have been able to shelter underground or in some sort of cavity and later emerge. One problem with this hypothesis is the differential survival of different mammalian clades in different areas. How can the infrared radiation scenario explain why eutherian mammals faired much better than metatherian mammals in North America, while in South America both groups radiated in the Early Tertiary? Similarly, why would the clearly aquatic sharks disappear completely from the Western Interior of North America?

Even more problematic are birds and lizards. Some number of bird clades clearly survived, yet there seems little basis to argue that they were crevice dwellers or even burrowers, even though the authors attempted to make just such an ad hoc argument. The authors claimed that lizards did well because they could hide. Higher-level lizard taxa do appear to have survived, but in the best sampled areas, such as eastern Montana, eight of 10 lizard species did not survive. These lizards ranged in adult size from a few centimeters to perhaps three meters. Global infrared radiation does a very poor job of explaining this quite severe local or regional extinction of lizards.

This scenario becomes even more unsatisfactory in the marine realm because it fails to explain victims as disparate as nannoplankton, planktonic foraminifera, ammonites, and marine reptiles versus survivors as disparate as benthic foraminifera, sponges, corals, lophophorates, and echinoderms.

Robertson and his colleagues also claimed that the intense heat ignited fires where there was sufficient combustible material. They argue that the intense infrared radiation ignited fires in various places almost simultaneously, and on land the fires would be especially intense because the radiation coming from the entire sky added heat as the fires burned. The idea of such global wildfires is, however, problematic based both on the rock and fossil records. In 2003 Claire Belcher and her colleagues indicated that lowered levels of charcoal at the K/T boundary, especially compared to horizons earlier and later, points to there not having been global wildfires. In 2004b, Robertson and his colleagues responded that the reduced level of charcoal at the K/T boundary indicates that the fires were so intense that the charcoal was consumed. Belcher and colleagues responded in the same year that if there had been complete removal of organic material in a global wildfire it should be seen in the sedimentological record, but it is not.[57] It must also be noted that a global wildfire as an extinction mechanism is probably far too severe in its supposed effects to have given us the differential pattern of vertebrate extinctions.

What is likely the final blow to the idea of global wildfire caused by thermal radiation and possibly the thermal radiation extinction model itself was published in late 2009.[58] The glaring error in the model of Robertson and his colleagues was that they failed to consider the shielding effects of the ejecta particles as they settled toward the ground. This self-shielding, which would increase as more ejecta fell, would have prevented global wildfires and the global killing of unprotected animals and plants. Local wildfires may have occurred, as well as the local extirpation of unshielded animals, including nonavian dinosaurs. On physical as well as biological grounds, the thermal radiation extinction model was at best a local or regionalized phenomenon.[59]

In 2005, Malcolm McKenna, one of my intellectual mentors and a coauthor on the Robertson and colleague paper, e-mailed the following: "Regarding non-reworked nonavian dinosaurs, our paper set a limit that, were they [nonavian dinosaurs] still around at that time [of the K/T boundary], they would not have been able to pass [the boundary]. Therefore, finding a non-reworked nonavian dinosaur in the Paleocene (or any other time later) . . . would falsify our story. This opens the door to scientific enquiry."

What McKenna meant by "non-reworked nonavian dinosaurs in the Paleocene" are the remains of dinosaurs that lived and died within the Paleocene, not specimens of animals that had long been dead but have been reworked, usually by water, and then redeposited in Paleocene deposits. This is what I have termed the Zombie Effect, obviously referring to dinosaurs that seemed to have been alive in the Paleocene because they were found in Paleocene rocks, yet they had actually died in the Cretaceous and were exhumed and redeposited in Paleocene sediments.[60]

Thus, according to Malcolm, finding a non-reworked nonavian dinosaur in Paleocene rocks would negate his and Robertson's hypothesis. I think, however, that he has created an unrealistic threshold. While I have noted my concerns with the infrared radiation hypothesis, I think Malcolm's statement sets the bar too far in the other direction. Even though he was a paleobiologist,[61] I think he underestimated the potential resilience of even those creatures that might not have been as fully protected from a blast as smaller individuals. I think this once again shows the danger of trying to establish an all-or-nothing setting for biotic systems when faced with dire physical changes, especially on a global scale. There are certainly problems with the infrared radiation hypothesis, but I think the evidence suggests that it may well have been a proximate factor, along with prolonged darkness following the K/T impact, but these hypotheses do not stand or fall on reports of a few non-reworked Paleocene dinosaurs of the nonavian sort.[62] Such hypotheses would only be falsified if it could be shown that substantial populations of nonavian dinosaurs survived somewhere in an unstudied or understudied part of the globe. Is there an earliest Tertiary version of Sir Arthur Conan Doyle's *The Lost World* yet to be recovered in the fossil record?—probably not.

MULTIPLE-CAUSE HYPOTHESES OF EXTINCTION
Volcanism

Another of the other undisputed events or, more correctly, series of events that is well established as having occurred across the K/T boundary is the emplacement of the Deccan Traps on the Indian subcontinent. Most recently these events have been championed by Gerta Keller and her colleagues, usually as *the* cause rather than one of a number of causes. In one of their more recent studies they found that something like 80% of the Deccan volcanism took place in about 800,000 years spanning the K/T boundary. As they note, because it has been difficult to determine when within this interval of major eruptions mass extinctions occurred, it has also been difficult to model how the eruptions could have caused major extinctions at the K/T boundary. This seems a fair assessment, no matter how one views the eruptions as a possible agent of mass extinction. The results of this particular study suggested that volcanism may have played a critical role in the K/T mass extinctions and may even have delayed biotic recovery. They reached these conclusions based upon the study of sediments and fossils from four-to-nine-meter-thick intertrappean[63] outcrops in four regions in southeastern India.[64]

Unfortunately, even this most recent study is not very forthcoming in explaining how such massive eruptions cause extinctions. As the authors note, "A possible cause-effect relationship between mass extinctions and volcanism has largely been inferred to date from their overall correspondence and the potential effects of volatile fluxes on the global environment."[65] Volatile fluxes, as the name suggests, refer to the gases, usually sulfur dioxide, released by volcanic activity. Most importantly, these gases and possibly very fine particulate matter lofted into the atmosphere may have helped cause or have triggered the cooling phase seen both in the terrestrial and marine records in the last few hundred thousand years of the Cretaceous. It is hard to imagine how this would have in and of itself caused the K/T mass extinctions, but it would most certainly have been a major contributing factor in environmental instability.

Finally, lest one have the impression that the Deccan Traps are just one more series of volcanic eruptions, these were the greatest prolonged eruptions during at least the past 100 million years, if not further back in time. These were generally not violent, such as the eruption of Mount Saint Helens; rather, enormous volumes of basaltic material erupted along great fissures in the earth. One of the earlier advocates of volcanism-based K/T extinction, Vincent Courtillot, noted that when they first erupted, the flows originally might have covered about two million square kilometers, with a volume of almost 1.5 million cubic kilometers.[66] For comparison, this is enough volcanic material to cover both Alaska and Texas to a depth of over 600 meters.[67] More recently, work attributed to Courtillot notes that the Pinatubo eruption in 1991 put about 0.017 billion tons of sulfur dioxide into the atmosphere.[68] This cooled Earth's climate for several years. By comparison, the Chicxulub impact is argued to have put between 50 billion to 500 billion tons of sulfur dioxide in the air. The entire Deccan traps supposedly spewed an amazing 10,000 billion tons into the atmosphere.[69] Although this certainly occurred over a longer interval of time, the amount is staggering, if true. Little wonder the latest Cretaceous world cooled down. There is, however, at least a potential problem with this scenario in that several lines of evidence point to a warming trend some 500,000 years before the rapid cooling trend leading to the K/T boundary. The above sources do not explain this warming trend, which is discussed later.

Marine Regression, Habitat Fragmentation, and a Cooling Climate

In the summer of 2008 I received a press release from the National Science Foundation titled "The Mystery of Mass Extinctions Is No Longer Murky." It said that "a new study, published online June 15, 2008, in the journal *Nature*, suggests that it is the ocean, and in particular the epic ebbs and flows of sea level and sediment over the course of geologic time, that is the primary cause of the world's periodic mass extinctions over the past 500 million years."[70]

My first response was a double take. I, as well as many before and after, have been arguing the same thing for many years. The paper in question, by Shanan Peters, appeared in print in late July. The author, unlike the National Science Foundation press release, made it very clear that it has been known for many years that changes in sea level coincide with large biotic changes. More importantly, Peters had found that the selectivity of extinctions of marine invertebrates was related to the environments in which they had lived, probably most notably by the replacement of carbonate sediments by terrigenous clastics in marine shelf environments.[71] In carbonate environments the rain of shells of microscopic organisms form an ooze on the ocean floor that may in time become limestone, while *terrigenous clastics* refer to the very fine-grained muds and sands up to the much rarer large chunks of material that are carried into the ocean by streams and storms.

The amount of such clastic material derived from the land increases during the major intervals when shallow epicontinental seas depart the land in the process known as regression. Transgression is the opposite phenomenon, when shallow seas invade lower-lying land. This is of course one of the great worries of global warming, which will certainly raise sea levels around the world. The transgressions and regressions of large parts of continents is most likely a byproduct of continental drift. One idea is that when places such as the mid-Atlantic Ridge, which separates major tectonic plates, is most active, the bulging upward of the ridge forces shallow seas on to the land. Except, possibly, for Hudson Bay, the Persian Gulf, and the Red Sea, we have no such epicontinental seas today.

Depending on who is doing the counting, there are usually five mass extinctions generally recognized during the Phanerozoic Eon, which commenced some 545 million years ago (fig. 5.1). These are usually referred to as the end-Ordovician (444 million years ago), Late Devonian (starting 375 million years ago), end-Permian (251 million years ago), end-Triassic (200 million years ago), and end-Cretaceous (or K/T, 66 million years ago) mass extinctions.[72] Of these, Peters singled out the end-Ordovician, Late Devonian, and Late Permian mass extinctions, as well as recognizing a sixth in the mid-Carboniferous, as times when there was a great loss of carbonate environments. After the end of the Permian extinction events he found too few appropriate carbonates to reliably cal-

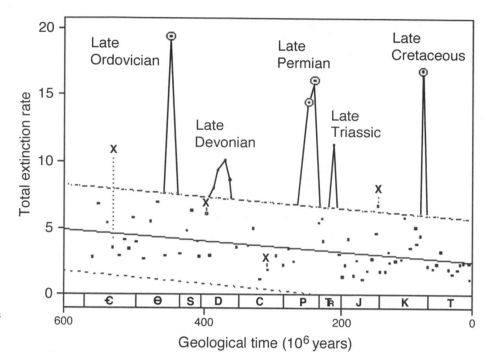

Fig. 5.1. The widely recognized five mass extinctions during the Phanerozoic Eon. Modified after Archibald 1996.

culate the loss of carbonate environments for the last 200 million years. These post-Permian times include the end-Triassic and end-Cretaceous (or K/T) mass extinctions. Nevertheless, these later two mass extinctions, like those before, occurred during or at the end of a time of massive regression of epicontinental seas.

Peters concluded that a very probable "hypothesis is that both evolutionary fauna turnover and shelf sedimentation share a common cause. The most plausible mechanism is sea-level change and the resultant expansion and contraction of epicontinental seas, phenomena which are intimately linked to tectonics via crustal uplift and subsidence, and to global climate via temperature and continental ice volume." He continued, "these results do provide a substantial measure of support for the intriguing and long-standing hypothesis that changes in the areas of unique epicontinental sea habitats, as well as correlated environmental effects, have consistently influenced rates of extinction, extinction selectivity, and the shifting composition of the marine biota during the Phanerozoic eon."[73] I could not have said it better myself, and while Peters was working on marine organisms and environments, the result is the same for plants and animals inhabiting the low coastal plains that abutted these shallow epicontinental seas—a mass extinction.

If Peters did not originate the connection between mass extinctions and marine regressions, who did? I know that my first introduction to the idea for this correlation was hearing from and reading about Norman Newell's emphasis on mass extinctions and their causes while I was a graduate student in the early 1970s. In the late 50s and 60s Newell was trying to convince the geological community of the existence of mass extinctions. One of his figures that most impressed

me showed the percentage of first and last appearances of animal families through geologic time (fig. 5.2). With very minor tweaking, his figure showed the five mass extinctions still recognized today. This is the precursor to a famous figure by David Raup and Jack Sepkoski showing the mass extinctions.[74]

Newell entertained various catastrophic and gradual causes for these mass extinctions, but one of his pets was the correlation of mass extinctions and marine regressions.[75] As Newell and others have pointed out, the idea of episodes of mass extinction is not a new idea, and neither is the idea that sea-level changes may be a culprit.[76]

The earliest prominent scientist to argue for mass extinctions was Baron George Cuvier. He is often presented as a catastrophist who was later overshadowed by Lyellian uniformitarianism.[77] With the advent of the K/T boundary impact, however, Cuvier was "resuscitated" as the champion of a neocatastrophism. What is usually not mentioned is that in Cuvier's discussion of causes of his "revolutions," a common topic is what we would call change in sea level. This is nowhere better seen than in two sections in his introduction on the fossil bones of quadrupeds titled "Discourse on the Revolutionary Upheavals on the Surface of the Globe and on the Changes Which They Have Produced."[78] These two sections are "Proofs that these revolutions have been numerous" and "Proofs that these revolutions have been sudden."

In the first section Cuvier writes,

Finally, we say that if we examine with even greater care the remains of these organic creatures, we come to discover in the middle of the marine strata, even the most ancient ones, layers full of animal or vegetable products from land and

Percentage of total families

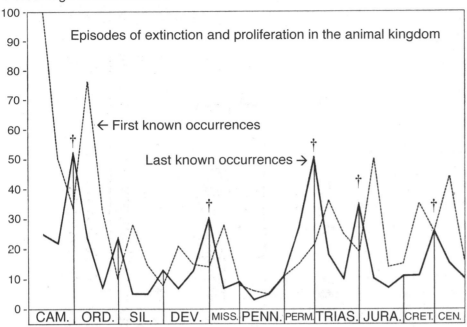

Fig. 5.2. Newell's 1967 diagram (text increased in size) showing five mass extinctions. From Newell 1967, with permission of the Geological Society of America.

fresh water. In the most recent layers (i.e., the ones closest to the surface) there are some where land animals are buried under masses of marine creatures. Thus, not only did the different catastrophes which moved the layers gradually make the various parts of our continent rise up from the bottom of the sea and reduce the size of the sea basin; but this basin has been moved in several directions. Often the regions converted into dry land have been covered again by the seas, whether they have sunk or the waters have been carried above them. . . . The changes in the heights of the oceans did not therefore consist only in one withdrawal more or less gradual, more or less universal. It was a matter of a succession of various eruptions and retreats. The result of these has definitely been, however, a general lowering of the sea level.

And in the second section:

But it is also really important to note that these eruptions and repeated retreats were not at all slow and did not all take place gradually. On the contrary, most of the catastrophes which brought them on have been sudden. . . . The rending, rearranging, and overturning of more ancient layers leave no doubt that sudden and violent causes placed them in the state in which we see them. The very force of the movements which the bodies of water experienced is still attested to by the mountain of remains and rounded pebbles interposed in many places between the solid layers. Thus, life on this earth has often been disturbed by dreadful events. Innumerable living creatures have been victims of these catastrophes. Some inhabitants of dry land have seen themselves swallowed up by floods; others living in the

ocean depths when the bottom of the sea was lifted up suddenly were placed on dry land. Their very races were extinguished for ever, leaving behind nothing in the world but some hardly recognizable debris for the natural scientist.

In Cuvier's words we can see that he placed great store in the effects of waxing and waning water, sometimes but not always at a catastrophic pace. While he does not name it as such, it is clear that at least some of his catastrophic events in the second passage involve the consequences of mountain building. Charles Lyell, often referred to as the father of modern geology, also considered changes of sea level of profound importance to Earth processes. Unlike Cuvier, however, Lyell's brand of uniformitarianism is often confusedly associated with the pace of change as always being slow and stately. In reality both Cuvier and Lyell argued for rapid as well as slow changes in the history. What some impactor theorists fail to understand in their newfound championing of Cuvier as a catastrophist is that his ideas were more complex than has usually been appreciated. As Martin Rudwick has pointed out, although Cuvier was certainly enamored of catastrophic revolutions, not all of Cuvier's revolutions were sudden,[79] and Cuvier pointed to long periods of quiescence.

Even given all the evidence for a severe regression in the latest Cretaceous, there are those who remain skeptical that it could have played a role in the terminal Cretaceous extinctions. Some incredibly even deny that there was such a regression! A discussion of these views can be found in Doug Nichols and Kirk Johnson's recent book on what happened to plants at the K/T boundary.[80] Unfortunately, the authors misstate some of my ideas related to marine regression. First, they indicate that my "favored" explanation for the K/T extinctions is marine regression—not so. At least from the early 1990s onward, I have emphasized that the differential pattern of extinctions points to multiple causes for these extinctions.

Nichols and Johnson next claim, "The simple fact is that none of the major episodes of extinction during the time [referring to the Phanerozoic]—including at the K-T boundary—correlate with the major episodes of regression."[81] In fact, *all five mass extinctions correlate with marine regression*, as the study of Peters discussed above demonstrates. They note that sea-level changes in the Pleistocene were not accompanied by "major extinctions of land plants or animals." This is a proverbial straw man. First, no one claims that sea level changes in the Pleistocene represent a major transgressive-regressive episode. Second, no one, including me, has said that sea-level changes per se cause extinctions, nor have I said that at the K/T boundary they are the sole cause.

The authors then assert that the major regressions in western North America were complete three to four million years before the K/T boundary. They cite the Breien Member and the Cantapeta Tongue of the Hell Creek Formation as evidence that the sea "was readvancing into the North Dakota area in minor transgressive pulses" before the end of the Cretaceous. They then go on to write that the last transgression into the interior of North America was represented by the Cannonball Member of the Fort Union Formation in the middle Paleocene. Here is an example of two groups of researchers seeing the same evidence but coming to quite different conclusions. There are two pieces of information— one geologic and one paleontologic—showing these authors most likely to be wrong. I start with the geologic data.

If, as Nichols and Johnson claim, maximum regression occurred three to four million years before the K/T boundary in this the northern Western Interior and the seas began to return in this time frame, continuing into the Paleocene, we should see this in the rock record. That is, there should be a sequence of marine rocks in this region transgressing the K/T boundary on into the Paleocene. This would support claims of transgression during this time. We do not find any such sequence. Instead, we find what the authors describe and show in their figure 6.2 (see fig. 5.3).

The section from bottom to top is a classic regressive phase from the marine Bearpaw Shale, to the near-shore Fox Hills Sandstone, and is finally capped by the Hell Creek Formation. Various studies of the paleochannels show streams in the upper Hell Creek Formation flowing to the east and south, toward the regressing sea.[82] The Breien Member and the Cantapeta Tongue of the Hell Creek Formation represent minor pulses before the seas leave the area.[83] In no instance in this region do Paleocene marine beds lie directly on marine Cretaceous beds. Rather, the terrestrial Paleocene deposits of the Fort Union rest directly on the terrestrial Hell Creek Formation.[84] This shows that the epicontinental sea

had left the Western Interior by the end of the Cretaceous. The sea did start to reinvade the southern interior of North America, possibly starting about one million years before the end of the Cretaceous, but continuous K/T boundary marine deposits along the seaway are manifest only in places much further to the south, such as the well-known Brazos River section in Texas,[85] although lesser-known marine K/T boundary sections have been reported as far north as southeastern Missouri (fig. 5.4).[86] As Nichols and Johnson argue and the rock record shows, the last gasp of the shallow seas returned to the interior of North America as the Cannonball Formation a few million years later, in mid Paleocene times, before departing for good.

The paleontologic evidence showing that the shallow seas departed the Western Interior by the end of the Cretaceous comes from sharks and their relatives. This might seem surprising, given that when one thinks of Late Cretaceous terrestrial vertebrates, nonavian dinosaurs, turtles, and maybe even mammals come to mind, but sharks? One does not normally think of sharks as living in freshwater, but as noted in Chapter 4, they do. Today, some sharks, notably bull sharks and some stingrays,[87] can be found thousands of kilometers inland from the coast in the Amazon basin. These are the exceptions. Although a number of other sharks and relatives are found hundreds of kilometers inland, they still have ties to and breed in marine conditions. This means that if we find them in the freshwater deposits, there must be marine conditions relatively close by. They are the "canaries in the coal mine," indicating the presence or absence of nearby marine conditions. I must also point out that amphibians, especially salamanders, are present in both Cretaceous and Paleocene terrestrial deposits in the Western Interior, which is a clear indication that the deposits in question were from a freshwater and not a marine source. Amphibians in general have a low tolerance for saline water.

In the Late Cretaceous Hell Creek Formation sharks and relatives are known, while in the Early Paleocene beds known as the Tullock Formation, or Tullock Member of the Fort Union Formation, no such species are found. They once again occur in mid Paleocene beds, notably the Cannonball Formation, noted above.[88] There is always a danger of putting too much stock in negative evidence, but the Early Paleocene Tullock and its equivalents have been well sampled. The sharks and relatives that were present in the streams of the Hell Creek Formation west of the receding sea during the Late Cretaceous were gone before or by the Early Paleocene, when the Tullock and equivalents were being deposited.[89] The connections to the shallow seas had been severed. The seas returned slightly later as the Cannonball Sea. Except for one species, these sharks and relatives are different from those in the Late Cretaceous.[90] Because there is an unquestionable gap of perhaps one million years in the shark record we do not know whether this change in the shark fauna was normal-paced evolution and extinction, a result of events at

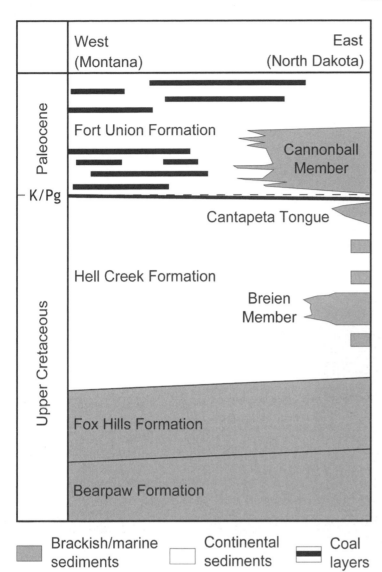

West
(Montana)

East
(North Dakota)

Paleocene

Fort Union Formation

Cannonball
Member

K/Pg

Cantapeta Tongue

Hell Creek Formation

Breien
Member

Upper Cretaceous

Fox Hills Formation

Bearpaw Formation

Brackish/marine
sediments

Continental
sediments

Coal
layers

Fig. 5.3. Schematic geologic cross-section from Montana (west) to North Dakota (east). Note that nowhere do marine Paleocene rocks rest directly on marine Cretaceous rocks. The implication of earlier Paleocene marine rocks is not correct. Marine rocks are not known until late Early (or mid) Paleocene. This version, modified and provided by Scholz, has the acronym K/Pg rather than the K/T boundary. From Scholz and Hartman 2007.

the K/T boundary, or some combination of the two. Whatever the cause, the pattern of sharks and relatives along with what we know of the geology shows that, contra Nichols and Johnson, there was a regression from the Western Interior during the latest Cretaceous. Its biologic ramifications cannot be ignored.

Finally, there is the issue of what the global climate was doing leading up to the K/T boundary and the impact at Chicxulub. Was it business as usual until the fateful moment or were there changes in the air—literally? Based upon correlation between the terrestrial fossil plants and the paleoclimate data provided by miniscule marine foraminiferans, a global warming trend is indicated within the last one million years before the K/T boundary, with cooling before the boundary. Further, the fossil record of plants suggests continued relatively cool temperatures across the K/T boundary, with no indication of a major warming trend immediately after the boundary.[91] During the last 100 thousand years

of the Cretaceous, Greg Wilson notes dramatic changes in mammal and turtle paleocommunities,[92] which may well be correlated to this cooling trend. More recently Greg wrote that "within ~100 thousand years of the K-T boundary, nearly 75% of all mammalian species in local faunas went extinct. A recovery of pre K-T species richness and morphological disparity rapidly occurred within the first ~200 thousand years of the Paleocene fueled by immigration and in situ evolution."[93]

There is an interesting little dinoflagellate, or unicellular flagellum-bearing protist, *Manumiella seelandica,* which occurs in great abundance in marine sediments just prior to the K/T boundary and is correlated with an interval of global sea-level regression and precisely with oxygen isotope evidence for cooling. As noted, within the last million years of the Cretaceous there was a period of global warmth, but a cooling period began within at least tens of thousands of years before the close of the Cretaceous Period. This helped

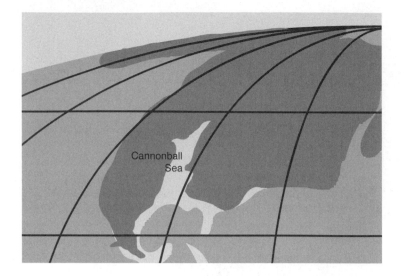

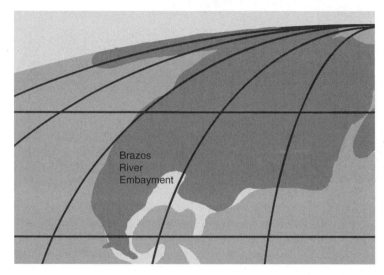

Fig. 5.4. North American paleobiogeography near the mid Paleocene *(above)* and the K/T boundary *(below)*. In part after Smith, Smith, and Funnell 1994 and Gale 2006.

produce this global assemblage of dinoflagellates, such as *Manumiella seelandica.*[94]

Thus, evidence from terrestrial plants and vertebrates as well as from marine microorganisms indicates warming and then cooling in the last tens or hundreds of thousands of years just preceding the K/T boundary. Some recent evidence suggests that it may relate to changes in plant communities leading up to the K/T boundary. In northeastern Montana, Anna Thompson and her colleagues used fine-scale stratigraphic sampling and found that the number of taxa of dicot angiosperm pollen dropped from 44 to 11 in the uppermost 3.5 m of Cretaceous sediments. While this may well be a local extirpation, data from other areas argue that the latest Cretaceous plant communities were shifting from more homogeneous to more heterogeneous. The authors of this study suggest that vertebrate lineages were rendered more vulnerable to impact-induced environmental disturbance. I agree with their assessment that this combination of factors better explains differential vertebrate extinctions and surviv-

als in response to the K/T boundary event.[95] Global regression of the shallow epicontinental seas was increasing the instability in the shrinking coastal environments. The impact of a large extraterrestrial object at the end of the Cretaceous was the tipping point to this deteriorating ecological scene.

Impact, One of a Number of Causes

I have already discussed in detail the issues surrounding the impact of an extraterrestrial object. As I hope is also clear, I regard the effects of darkened skies and some sort of infrared radiation as the most likely factors that can be supported by the vertebrate fossil record. It is still clear, however, that while likely necessary, these proximate extinction causes resulting from the K/T boundary impact are not sufficient to explain all that happened at the time.

There is no single pattern of extinction and survival at the K/T boundary. The great phylogenetic and ecological differences between who survived and who did not clearly points to multiple causes. Together, volcanism, marine regression, and an extraterrestrial impact do explain this differential pattern of survival. Certainly, almost everything that could go wrong leading up to the K/T boundary did go wrong.

It must be reiterated that for the other four mass extinctions in the Phanerozoic Eon, the last 550 million years, there is no evidence of an extraterrestrial impact. Also, as discussed earlier, none of the other major impacts during the Phanerozoic resulted in mass extinctions. In stark contrast, all five major mass extinctions, including that at the K/T boundary, are correlated with major global marine regressions.

The best-supported scenario for the K/T extinctions is one of multiple causes: volcanism, marine regression, and asteroid impact. It is a history of more than one kind of biotic stress leading to extinction.[96]

One must start at least 10 million years before the boundary to understand what transpired. Because the record of vertebrate change during this time is largely limited to the Western Interior, this scenario must by necessity be similarly limited. How much more widely it can be applied remains a solvable paleontologic puzzle.

Approximately 75 million years ago an open plain with scattered trees existed on the eastern shore of Laramidia (Western Interior of North America). Rivers of moderate size meandered through the landscape. In the distance, the plains were lapped by a shallow sea, which stretched to the horizon. The plains were dominated by vast herds of several species of duck-billed and horned dinosaurs, much as the Serengeti Plain of Africa today is dominated by large grazing mammals. Other ornithischian dinosaurs and the infrequent meat-eating theropods crossed the landscape. The streams were populated by various species of turtles, amphibians, crocodilians, and fish, including the occasional skate or shark swimming up from the nearby shallow sea. It is daytime; mouse- and rat-sized mammals are asleep in their dens.

Fast forward to about 66 million years ago. The shallow seaway has begun to slip rapidly away to the southeast. As the exiting seaways reach lower-lying, flatter terrain, the rate of exodus quickens, with the final stages of withdrawal occurring in, at most, tens of thousands of years. So too, the great herds of duck-billed and horned dinosaurs have diminished. As the dwindling refugia of low coastal plains rapidly decreases, first one, then another of the species dwindles, until the great herds are reduced to at most two or three remaining species, much like the herds of bison that once roamed North America. Nonavian dinosaurs, like large vertebrates everywhere and at every time, are the first to experience biotic stresses leading to decline and disappearance. The plants are changing too, from more homogeneous stands stretching as far as the eye can see to more heterogeneous stands in the dwindling plains. The skies are darker than they had been, caused by the faraway eruption of the Deccan basalt flows. This brings a cooling climate, which may have put a burden on the more warmth-loving species of plant and animal.

We do not know what was happening to vertebrates in more inland areas just before the K/T boundary, as few such areas are well preserved and none has been studied. The coastal plains nonavian dinosaurs certainly were capable of migrating from one shrinking coastal habitat to another, but finally even this could not stop further declines in population sizes—just like the relentless encroachments of increasing human populations that decimated the American bison and are causing many biotas to shrink today. Other large vertebrates suffered. The Komodo Dragon–sized lizards and the single exclusively terrestrial turtle, *Boremys*, also experienced declines. Populations of smaller terrestrial vertebrates were also declining, but because of shorter life spans and faster turnover rates, they adapted more quickly to the environmental stresses caused by the loss and fragmentation of the coastal plains.

Marsupials had flourished for some 25 million years in North America. Newly emerging land bridges appeared as the seas retreated. Invaders appeared. In North America there were newly arriving diminutive archaic ungulates, possibly from Asia, to join the few that had reached North America during the Late Cretaceous. In the Western Interior, at least, they outcompeted the marsupials for dwindling resources. In South America events were different. Both groups of mammals appeared in South America at or soon after the K/T boundary, but here they divided the guilds, with marsupials becoming the carnivores and the ungulates the herbivores. This coevolutionary arrangement lasted for almost 50 million years in South America, with only an infusion of rodents and primates from the outside world (probably Africa), until about five million years ago, when a trickle and then a flood of mammals were exchanged between North and South America.[97]

Unlike the terrestrial vertebrates, freshwater species faced far less stress, especially because the size of their habitat was at least holding its own as the lengthening streams followed the retreating seas. Not all aquatic vertebrates fared so well. With the loss of close ties to the seas in areas like eastern Montana, sharks and skates ventured into the rivers in the area less and less frequently, as the distance to the sea expanded from tens to thousands of miles, eventually reaching southern Missouri and even Texas, to the south.

Plants and near-shore species also showed added stresses as their respective habitats shrank. Certainly, some species must have done fine as new habitats were formed as the seas regressed. As with vertebrates, however, we do not have any clear record of these environments away from the coastal areas.

The Deccan basalts continued to spew forth, adding further stresses. The added particulate matter in the atmosphere continued to cool and dry some areas of the globe, probably at an accelerated pace.

Suddenly, a literally Earth-shattering event magnified the

differences between the "have" and "have not" species. A very large extraterrestrial object struck what today we call the Yucatan Peninsula. Fine material and aerosols injected into the upper atmosphere formed a cover of darkness, shielding the sun to the point that photosynthesis ceased or diminished for many weeks, depending upon the location. A rain of material in at least some parts of the globe created a searing blanket of infrared radiation, which roasted any exposed plants or animals. This alone spelled doom for some species.

The effects of darkness were especially acute at lower latitudes and closer to the impact, for example, in North America. Plants unaccustomed to lower-light regimes caused by seasonal changes in the Sun's position were especially hard hit. Higher latitude plants, accustomed to seasonally lower-light regimes, survived much better, as did the animals that fed upon them. The effects on higher-latitude plants and animals were tempered by which season they were experiencing when the impact occurred. Extinction rates for coastal plants in North America soared because of the cumulative effects of continued habitat loss, drought, and loss of sunlight and possibly because of being closer to the impact site.

Except for the sharks and relatives, who had already departed or become extinct as the seas regressed, all ectothermic, aquatic vertebrate species (bony fishes, amphibians, turtles, champsosaurs, and crocodilians) weathered the impact quite well in their still-flourishing freshwater habitats. Their watery cover also did a good job of shielding them from infrared radiation.

With the added loss of more plant species and the reduction of biomass that the impact brought to the already highly stressed ecosystem on land, other vertebrate species rapidly succumbed. Most notable were the last of the large herbivorous nonavian dinosaurs. The remaining predaceous, nonavian dinosaurs followed soon thereafter, with the larger species disappearing first. In some places on the globe the great saurians may have lingered a while longer, but finally, for the first time in more than 150 million years, no large land vertebrates graced the Earth. The landscape was open and waiting for evolution's next gambit—the rise of the mammals.

6

After the Impact
Modern Mammals, When and Whence

AS IS OFTEN MORE TRUE than not in science, seemingly unrelated or at least distantly related topics are later found to be inexorably entwined. This is the theme of this chapter. The extinction of nonavian dinosaurs is generally, and I think rightly, given as the major cause of the explosion of mammals into many lineages and amazing ecological diversity. But what was the pace of this diversification? Did it actually begin before nonavian dinosaur extinction, as molecules argue, or only after the reptilian behemoths were gone, maybe long gone, as the fossil record argues? More succinctly put, does the extinction of nonavian dinosaurs have anything to do with the pace of mammalian evolution—are they inexorably related? My answer is absolutely yes.

LIVING MAMMALS AND THEIR EXTINCT RELATIVES

Recall from Chapter 3 that, based on our current understanding, the crown group (living mammals) within Mammaliaformes, which we call Mammalia, is composed of two major clades, Prototheria and Theria. For a phylogeny of these relationships, see figure 3.1.[1] Extant prototherians include only five species within the order Monotremata—echidnas and the platypus—today found only in Australia and New Guinea. Theria includes Metatheria and Eutheria, whose crown clades are Marsupialia and Placentalia, respectively. Marsupialia has some 237 species in Australia, 94 species in South and Central America, and only one in North America (our opossum). Placentalia today has 5,080 species found on every continent, in every ocean, and occupying the air on many parts of the globe.[2] Only in Australia are placentals surpassed by marsupials in numbers of species.[3] Except in South America and Australia, stem eutherians and their crown placentals have dominated both ecologically and in terms of numbers of species at least since shortly after the K/T boundary, if not before.

I briefly recount the molecular and morphological evidence regarding the geographic history and the timing of the origin for the crown clades Monotremata and Marsupialia, but the majority of this chapter is devoted to the controversies surrounding when and where the vastly more species-rich Placentalia arose. Interestingly, with a few exceptions, there are far fewer disagreements concerning the relationships of major clades of placentals than there were just a few years ago. Molecules and morphology are converging on the major relationships within Mammalia. This is not so, however, when we turn to the questions of when and where the major clades of placentals arose.

PROTOTHERIANS AND MONOTREMES

Although fossil and living echidnas have no vestige of teeth, probably because the lower jaw is so atrophied, the platypus does have embryonic teeth, which are reabsorbed before eruption. In part because of the morphology of these embryonic teeth and some of the other peculiar aspects of monotreme anatomy, the group can be traced back to the Early Cretaceous, some 105 to 121 million years ago in Australia. Some of these specimens are fossilized as beautifully iridescent opal. From that time on, monotremes appear sporadically in the fossil record of Australia and today occur in Australia and New Guinea.

Then, in 1992, a single but unmistakable tooth of a monotreme, probably of an ornithorhynchid, or platypus, was reported from the Paleocene (about 60 million years ago) of Argentina.[4] Although a very nice discovery, it was not a great surprise because Australia was once part of the greater supercontinent of Gondwana, which began to break apart in the Jurassic, some 165 million years ago. The process accelerated in the Cretaceous, first with Africa, then India, and finally Australia moving north. India had certainly crashed into south Asia by the Eocene, creating massive mountain ranges such as the Himalayas, but the collision could have begun as early as the Late Cretaceous. Australia (plus New Guinea) will do the same thing to Southeast Asia in the not-too-distant geological future.

Other mammalian groups are known from the Jurassic and Cretaceous of Gondwana, but it is noteworthy that therians (metatherians and eutherians) are not known with certainty from any southern continent until the latest Cretaceous.[5] This point will be very important in the biogeographic discussion at the end of this chapter.

Based on the fossil record, we have the earliest protetherian and possibly crown montremes by at most 121 million years ago. Figure 3.1 can serve as a refresher for this as well as for the other events in mammaliaform evolution I discuss. Fossil metatherians and eutherians have been argued to be present from about 125 million years ago, and eutherians are known with more certainty 105 million years ago, while molecular evidence suggests 148 million years ago for the split between these to clades.[6] Accepting these dates and the above identifications means that protetherians and therians must have shared a common ancestor no later than 125-plus million years ago and almost certainly older. Fossils from between about 150 to 180 million years ago, but possibly as old as the Late Triassic (some 200 million years ago), have been implicated as stem to Theria.[7] Our understanding of these early mammaliforms remains in flux. The most conservative assessment based on the fossil record is that the origin of Mammalia (and thus the stem Prototheria and stem Theria) could date to 200-plus million years ago (but could be as young as about 135 million years ago).

A study published in *Nature* in 2007 by Olaf Bininda-Emonds and colleagues, about which I will have much more to say later, placed the origin of Prototheria (including the crown Monotremata) at about 166 million years ago, based on molecular data. This would also mean the origin of their sister group, Theria, at the same time.[8]

What is the correct date for the split between Theria and Prototheria, and hence the origin of the crown clade Mammalia? In this instance the fossil record only provides a range of between 200 and 135 million years ago for the origin; thus, there is nothing to support or refute a molecularly based date of 166 million years ago for the origin of the crown group Mammalia.

CROWN THERIA—THE EARLIEST METATHERIANS AND EUTHERIANS

In papers in 2002 and 2003 unmistakable but thoroughly squashed mammal specimens from the 125-million-year-old Yixian beds of China were hailed as the earliest known eutherian and metatherian,[9] respectively. While these claims may well prove to be true, only-low resolution photographs of these creatures have to date been published. This means that all assessments of characters in these specimens are based on interpretive reconstructions of the authors.

Because metatherians and eutherians are each other's sister clade, like twins separated at birth, the two clades must have originated at the same time, whatever that time may be. While my analogy of twins is admittedly strained, it nevertheless emphasizes that the two clades arose at the same time; metatherians and their crown group, marsupials, are not more primitive ancestors of eutherians and their crown group, placentals. This means that if *either* of the Chinese fossils turns out to be what is claimed, metatherians and eutherians both arose some 125 million years ago. I do not take issue with this early date and would not be surprised if it could eventually be pushed back even further, maybe to 150 million years ago. Using the next oldest but more widely recognized metatherian or eutherian brings us to 105 million years ago and a fossil eutherian from Mongolia known as *Prokennalestes*.

METATHERIANS AND MARSUPIALS

Metatherians are known sporadically in Asia from about 90 million years ago, but their greatest diversity is in the Cretaceous of North America starting about 100-plus million years ago. Here we see a minor radiation with as many as 13 species by the end of the Cretaceous. Much more fragmentary material is now also turning up in Europe.[10] There are only a few Paleocene marsupials in Europe, but Eocene marsupials in Europe are quite common and diverse. In the Oligocene they decline, but the Miocene species *Amphiperatherium frequens* is common, as its name implies.[11] A few metatherians are also known in the early Tertiary of the Middle East and Africa, but it is in South America and Australia via Antarctica that the crown group of Metatheria—Marsupialia—flourishes.

The actual timing of the arrival of metatherians and possibly marsupials in South America and then Australia is not certain, but no metatherians are known from either continent with any certainty until shortly after the K/T boundary. In the case of South America, and to a much lesser extent Australia, this is not based on the lack of any pre-K/T mammalian fossils, as some nontherian mammals are known from these two continents. But these are not metatherians.

Metatherians, including marsupials, and eutherians begin to blossom at least five million years after the K/T boundary in South America, and they are known by well-preserved, sometimes complete skeletons at Tiupampa in Bolivia.[12] From then on in South America marsupials radiate into dog-sized forms known as borhyaenids, and even saber-toothed forms are known. While metatherians were the omnivores and carnivores, eutherians evolved into a clade of native South American herbivores. But by about three million years ago only the opossum line, Didelphidae, and two lesser groups, the shrew opossums (caenolestids) and the monito del monte (microbiotherids) are known. As noted earlier, these three living marsupial clades comprise some 94 living species, mostly didelphids, with the ecological diversity showing a similar decline today compared to earlier in the Tertiary. The eutherian herbivores faced a similar decline, but, unlike the marsupials, the former became extinct.

Then, by at least 55 million years ago, marsupials were in Australia, already showing the inklings of what was to come for marsupial diversity on that continent. The fossil site Tingamarra yields the earliest known marsupials from Australia, but their possible relationships to living groups of Australian marsupials remains uncertain.[13] It will be no surprise if some of these taxa can be linked to modern groups of Australian marsupials.

To recap what we know of the timing of metatherian and marsupial evolution based on the fossil record, metatherians split from eutherians sometime between 105 and 125 million years ago, on a northern continent, possibly Asia, although this could have happened earlier. The crown metatherian group Marsupialia may have arisen between 70 and 66 million years ago, again possibly in Asia. Metatherians (including marsupials) reached and began radiating in South America near the K/T boundary some 66 million years ago. By at least 55 million years ago they finally reached Australia, where they underwent another great radiation. I should note that a few marsupials are known from Antarctica, sealing this as the route taken from South America to Australia.

How does this fossil record agree with what has been determined using molecular data? As noted previously, the molecular split between Metatheria and Eutheria is suggested as 148 million years ago compared to 105 to 125 million years ago based on fossils. The origin of the crown Marsupialia, which is the split between didelphids and their relatives and all other marsupials, is gauged by molecules as being between 89 and 79 million years ago compared to between 70 to 66 million

years ago based on fossils. The first lineages of modern Australian marsupials appeared about 63 million years ago based on molecules, compared to 55 million years ago based on possible Australian ancestral forms.[14]

In all three of these splits, the molecular dates are older. In the first instance, for the origin of metatherians, the molecular dates are older by 23 million years or about 18% greater than the fossil date. In the second instance, the origin of Marsupialia, the molecular dates range between 23 and 9 million years, or between 35% and 13% older than those based on fossils. The radiation of Australian marsupials is eight million years or almost 15% older comparing fossils to molecules. I would have to say that none of these discrepancies is wildly out of line given that the fossil record for the first two splits is simply not robust enough to dispute or to agree with what we know from the molecules.

The third split, the radiation of Australian marsupials, given all the possible vagaries there are, is actually in quite remarkably close agreement between fossils and molecules. This is especially true when the biogeographic part of the story discussed above is added to the mix. The molecular analysis argues that Australian marsupials began their radiation only after reaching Australia some 63 million years ago. The 55-million-year-old Tingamarra marsupials, while not definitely aligned with any living clades of Australian marsupials, nevertheless show that marsupials had arrived and were posed to radiate by 55 million years ago, if not earlier.

EUTHERIANS AND PLACENTALS

Let us now turn to the radiation of Eutheria and their living crown, Placentalia, examining both the issues of when and whence they arose. We first start with the players in this evolutionary drama.

If one believes the press, placental relationships have been turned on their head by molecular studies. There is some truth to this, but it must be made clear where there have and have not been significant changes in our understanding of placental mammal relationships as a result of molecular studies. When molecularly based phylogenies are compared to traditional analyses, there is much less disparity than is usually portrayed. Figure 6.1 is based on several commonly cited molecularly based phylogenies.[15] The numbers and memberships of the orders of placental mammals are very similar for molecularly and morphologically based taxonomies. Sixteen of 18 traditionally based orders are also recognized in most molecular studies.

The major differences are Cetacea and Lipotyphla, or whales and insectivores. Recent fossil evidence indicated that Cetacea and Artiodactyla (even-toed ungulates—pigs, hippos, cows, giraffes, etc.) may be each other's closest living relatives. Molecules have gone further, embedding cetaceans within Artiodactyla as the nearest relatives of hippopotami and also sometimes recognizing an unnecessarily new name: *Cetartiodactyla*.[16] The order Insectivora has gone through suc-

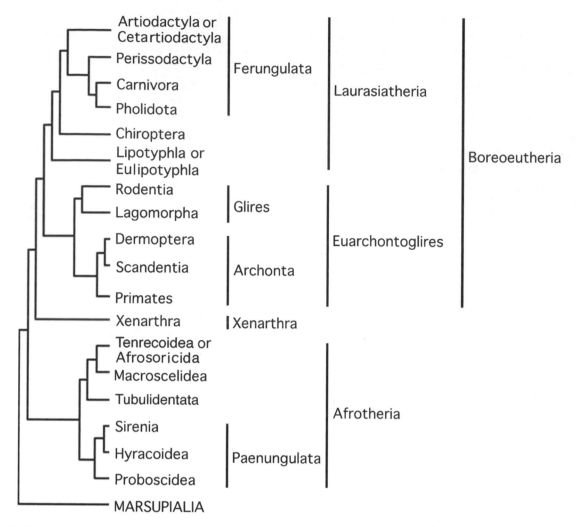

Fig. 6.1. Ordinal level phylogeny of placental mammals based on several commonly cited molecularly based phylogenies (see Chap. 6, n. 15).

cessive iterations over the years, first losing Menotyphla, which included Scandentia (tree shrews) are nearer our order Primates, and Macroscelidea (elephant shrews), which are neither elephants nor shrews. This left Lipotyphla representing Insectivora. More recently, molecular studies have pulled Tenrecidae from Lipotyphla, which has no common name other than *tenrecs*. They are African and Malagasy mammals that range from hedgehoglike mammals to otterlike forms. Also removed from Lipotyphla are the Chrysochloridae, meaning "golden pale green" referring the mammal's fur color. Tenrecidae and Chrysochloridae are now united in a clade called *Tenrecoidea* using the older name, or *Afrosoricida*, using a newer name. What are left of the insectivores are shrews, moles, and hedgehogs, along with a few Caribbean species referred to as Lipotyphla or Eulipotyphla.

The greatest disparities are found at the superordinal level, that is, clusters of placental orders, in which molecular studies have recognized some new and different combinations of orders. The four most commonly recognized superordinal clusters based on molecular data are Laurasiatheria, Eu-

archontoglires, Afrotheria, and Xenarthra.[17] Even here, older, anatomically defined groupings of orders, such as Glires and Paenungulata,[18] are identical to molecularly based taxa, or they are similar, such as Ferungulata and Archonta[19] (see fig. 6.1).

The common perception, derived almost exclusively from paleontology, has been that the 18 traditionally recognized orders of mammals and, to a lesser degree, their superordinal groupings did not make their appearance in the fossil record until about the first 5 to 15 million years after the K/T boundary. In the context of Earth history or even the shorter 200-million-year history of mammal evolution, this was an Explosive Model of radiation, a term Doug Deutschman, a colleague at San Diego State University, and I called this pattern in a 2001 paper (fig. 6.2).

Many, but certainly not all molecular biologists examining the same question came to a very different conclusion, namely, that most if not all modern orders of placental orders appeared slightly less than 100 million years ago, some 35 million years before the K/T boundary! Because it was being

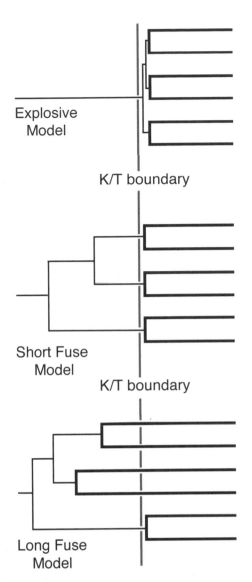

Molecular studies began tempering these claims of older Cretaceous ordinal appearances and began arguing that it was not the orders themselves but rather clusters of orders or superorders that began to appear in the Cretaceous. The fossils were suggestive of this possibility, and thus some paleontologists, including me, did find the idea tenable that there were Cretaceous superordinal clades predating the early Tertiary appearance of the orders. Deutschman and I accordingly called this the Long Fuse Model of ordinal appearances (see fig. 6.2). I discussed in Chapter 3 the hypothesis that others and I had argued that superordinal groupings of placentals existed in the Cretaceous. I have accepted the evidence of Wible and colleagues in their 2007 paper that as of now we have no definitive evidence of placentals in the Cretaceous, including superordinal clades. As Wible and his colleagues pointed out, however, they were not arguing that such placentals could not be in the Cretaceous but rather that their analyses do not show this based on the current fossil record.

A POST-K/T GARDEN OF EDEN

While the events leading up to and at the K/T boundary were not a picnic for the vast majority of plant or animal species, the fossil record soon after the extinctions shows that an evolutionary Garden of Eden of sorts prevailed for mammals for almost 10 million years.[21] This was a time of very rapid appearances of new mammalian species, one of the fastest if not the fastest for all of mammal evolution. In just the first few million years of the Tertiary, the number of genera and probably species as well jumped by a whopping factor of more than five! Once again, the best record of these events is in North America, so much of what I now discuss calls inordinately upon this area. I can, however, start with a global record taken from Doug Deutschman's and my 2001 study, discussed above. If we plot the global number of mammal genera from the Late Cretaceous to the present using these data, four patterns emerge (see fig. 6.3): (1) a dramatic surge in generic richness just after the K/T boundary, (2) reaches a peak in the Eocene, (3) is followed by a decline beginning in the Oligocene, (4) and is followed by another surge into the Recent.[22]

The first three patterns are quite real, but the fourth may not be. If I were to make an educated guess, I would say that because of our far greater, yet wholly incomplete knowledge of extant mammalian biotas, the apparently much greater present-day taxonomic richness is artifactual. If we could similarly sample older fossil mammal biotas, the entire graph would be pushed upward, with the Eocene peak at least matching or even surpassing today's richness. The dotted line in figure 6.3 suggests my speculation on this matter.

Let us then review the three patterns that are very likely real, working backward from the taxonomic decline beginning in the Oligocene. A comparison of the taxonomic richness curve to a plot of global climate change of the past 66 million years reveals that in the Oligocene the globe cooled

Fig. 6.2. Three models for the time of origin and diversification of super- or interordinal placental mammal clades (light lines) and ordinal placental mammal clades (heavy lines). Modified after Archibald and Deutschman 2001.

argued that the orders appeared relatively soon after the possible origin of the superorders around 100 million years ago, Deutschman and I termed this the Short Fuse Model of ordinal appearance (see fig. 6.2). Such early appearances of modern placental orders would have profound implications. For one thing, it implied that the fossil record simply could not be trusted. There must be many lineages and species of Cretaceous placental mammals out there that have gone undetected. While there is certainly truth in the general assertion that the fossil record can be sparse, chances that paleontologists had missed any possible trace of this earlier radiation of placental mammals seemed to border on the absurd—more on this later. The paleontological studies of Foote and colleagues,[20] as well as Deutschman's and mine, made forceful statistical arguments that this could not be the case.

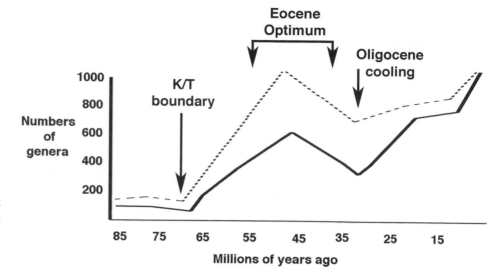

Fig. 6.3. Approximate global numbers of mammal genera from the Late Cretaceous through the Recent (solid line) and estimate of numbers assuming that the Eocene peak rather than the Recent corresponds to the interval of greatest diversity (dashed line). Data after Archibald and Deutschman 2001.

for a variety of cascading reasons—a circum-Antarctic current was fully formed, resulting in Antarctic glaciation accompanied by general global cooling and a decline in mammal species richness[23] (fig. 6.4). From then on into the Miocene, the geologic and paleontologic record shows a cooling and drying trend in North and South America. In South America, the Andes were continuing their dramatic rise, changing the landscape toward the cooler and drier climate of Patagonia and the Pampas. In Africa, the eastern forests gave way to today's savannahs, chasing both our ancestors and those of apes into the reducing forests. This is when we see the great concomitant appearance of larger grazing mammals across North America, Africa, Asia, and possibly elsewhere.

Stepping further back in time we reach a peak of mammalian diversity that, as I suggested above, might represent the interval of greatest mammalian diversity that Earth has ever seen. Climatically, this correlates well with the Eocene Optimum, probably the warmest interval in the past 66 million years. From what we can tell, the planet was more heavily forested with subtropical regions, where crocodilians and other warmth-loving creatures cavorted near the Arctic Circle—something they had not done even in the warmer parts of the Late Cretaceous.

Finally, moving still further back in time, we reach the K/T boundary, the time during which the steep climb to Eocene taxonomic richness began. As noted, in the first few million years of the Tertiary, species numbers jumped by an astounding factor of more than five. The vast majority of these are attributable to lineages that became extinct by at least the Oligocene, if not earlier, and thus did not contribute to the modern placental biota. From what can be determined, the global climate had been cooling quite considerably up to the K/T boundary,[24] after which it began its rapid ascent.

These patterns are in themselves quite interesting, especially when the corresponding pattern of climate change in figure 6.4 is matched to the taxonomic diversity curve in fig-

ure 6.3. This is not, however, the whole or, for our purposes here, the most interesting part of the story. To further examine this part of the story we need to look at another kind of pattern, which goes beyond the numbers and looks at who makes up these ups and downs in taxonomic richness. One of the better methods to show these changing patterns is a proportional mapping of major mammalian groups over the past 66 million years. There is not a reliable global compilation of this proportional pattern, but John Alroy has done this for terrestrial mammals of North America (fig. 6.5). Of the first three patterns in global mammalian diversity just discussed, the first two in the Cenozoic are clearly manifest in Alroy's graph, whereas the third is not.

At the K/T boundary one can see a precipitous drop in the number of species of the metatherians and rodentlike multituberculates,[25] neither of which recover their Cretaceous halcyon days in North America. While multituberculates do continue their downward trend, they nonetheless remained a major component especially of the earlier Paleocene mammal faunas. The newcomers are our old friends the condylarths, discussed in earlier chapters. As a refresher, this is the nonmonophyletic group of eutherian mammals that has been implicated in the origins of various placental orders, most notably of ungulate orders.[26]

As the Paleocene proceeds, the relative contribution of condylarths begins to wane, for two reasons. First, the "other" in figure 6.5 includes a host of eutherians, the vast majority of which do not belong to modern groups.[27] It does include some rather weird and wonderful forms, from dog-sized diggers (taeniodonts) to stilt-legged, rhinoceros-sized herbivores with daggerlike canines and multiple bony protuberances on their heads (dinoceratans). Second, we do see the first members of the three modern orders—Primates, Carnivora, and Lipotyphla ("Insectivora")—however, with a few possible exceptions, none of these newcomers belongs to modern clades within these orders; rather, they are stem

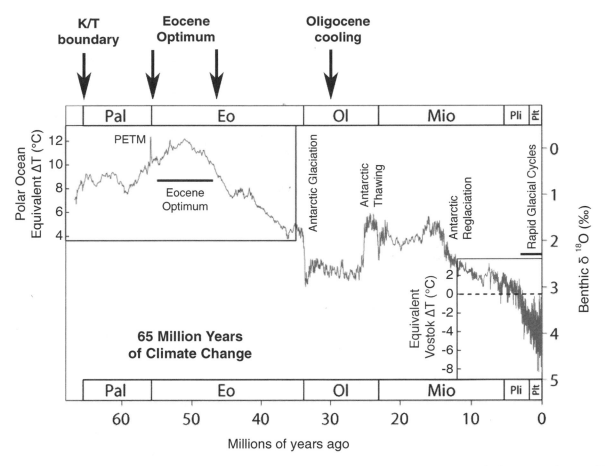

Fig. 6.4. Climatic change over the past 65 million years, indicating K/T boundary, PETM and Eocene Optimum, and Oligocene cooling. Modified from image created by Robert A. Rohde / Global Warming Art, www.globalwarmingart.com/wiki/Image:65_Myr_Climate_Change_Rev_png.

members not leading to modern species. Modern lineages do not appear until after the Paleocene Eocene Thermal Maximum (PETM), a major spike in temperature that heralds the Eocene Optimum (see fig. 6.5). As the graph shows, four additional extant orders of placental mammals—Perissodactyla, Artiodactyla, Lagomorpha, and Rodentia—appear after the Paleocene Eocene Thermal Maximum in North America. Rodentia and Lagomorpha actually appear slightly earlier, notably in Asia.

The third trend in global mammalian diversity, in this case decline, does not have as clear a signal (as the first two in Alroy's graph) for the simple reason that this is a proportional chart—it does not show total decline but only relative decline among orders or equivalent groups. These orders did decline in constituent taxonomic diversity, but their relative diversities remained similar through the remainder of the Cenozoic. By this time in North America (but not on other continents, notably, South America), the vast majority of mammals belonged to modern orders. Beginning in the Oligocene, there was the appearance noted earlier of larger, grazing mammals across most of the continents, but this is of course not shown in such a graph as this. Nevertheless, one pattern is a clear indicator of what was transpiring climatically. Note that the

primates, our order, all but disappeared from North America. This is completely in agreement with the cooling trend in North America and the accompanying loss of subtropical-to-tropical forests. North America had become inhospitable to our most ancient primate relatives until we dared show our faces again in North America perhaps tens of thousands of years ago. (The appearance of primates in South America is discussed later in the chapter.)

Let us return to the first two trends—the dramatic surge in generic richness just after the K/T boundary and the reaching of a taxonomic peak in the Eocene—as these two relate to how the K/T extinctions have inexorably molded mammalian evolution. While the diversity curve was ever upward from the K/T boundary to the Eocene peak, John Alroy's graph shows the appearance and notable expansion of a number of extant orders of placental mammals in North America in the early Eocene, a trend that was almost certainly nearly global in extent. The question for Mother Nature is, Why abandon the apparently successful mammalian faunas? Were these successful mammalian faunas dominated by condylarths for some 10 million years, only to be replaced by a whole new set of newfangled and as yet not thoroughly tested models? The likely catalyst, but probably not total cause for this significant

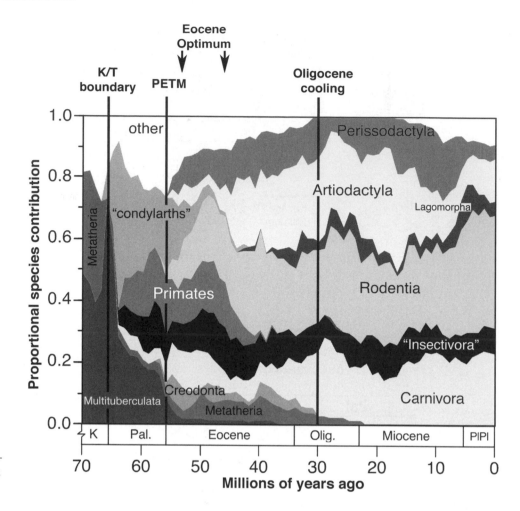

Fig. 6.5. Relative species diversities of major clades (mostly equivalent to orders) of North American mammals. Figure supplied by John Alroy, the Paleobiology Database.

faunal turnover, was the major global spike in temperature, the PETM, noted above. A major factor would have been an increase in selective pressures as more and more rather generalized condylarths vied for ecological space (fig. 6.6). This resulted in a shift from what has been termed a nonadaptive to an adaptive radiation. I am not a great fan of the term *nonadaptive radiation,* which conjures up a vision of poorly adapted species trying to compete and survive. Nonetheless, with this complaint duly noted, I use this term,[28] which has been explored in recent literature.

A *radiation* is, as the name implies, a flowering of many new forms. In an evolutionary context, the term signifies the appearance of a host of new species in a relatively short interval of time. The radiation is said to be nonadaptive if the myriad new species do not vary much one from another in form and ecological differentiation. On the other hand, if the radiation is adaptive, the new species rapidly differentiate morphologically and ecologically. I think that a very good case can be made for a shift from a nonadaptive radiation, which commenced at the K/T boundary and continued until near the beginning of the Eocene, when a significant adaptive radiation began, which brought us today's mammalian orders (see fig. 6.6). While the Eocene members of mod-

ern orders certainly did not vary as much in size as do modern members, the orders could already be recognized morphologically and, by implication, ecologically. In the Eocene even the extremes of today's mammals—whales and bats— are recognizable. Yet in the Paleocene, with a few exceptions,[29] such as the earliest members of some living placental orders, the condylarth-dominated mammalian biotas do not show nearly the same morphological and ecological diversity as do the Eocene mammals—a great amount of speciation with relatively less diversification, or what is termed a nonadaptive radiation. The cause of this pattern, which I term the Release-Restraint Selection Model, traces to the K/T boundary. *Release* refers to the relaxed selection for eutherians shortly after the K/T boundary, and *restraint* refers to the greater selective pressures that later returned.

With the extinction of the nonavian dinosaurs, coupled with the devastating effects of an extraterrestrial impact, the ecological stage was bare of large land vertebrates for the first time since the early Triassic, some 185 million years earlier. While there is some indication that a few species of condylarths may have been around just prior to the end of the Cretaceous,[30] they did not begin to appear in any great numbers until the early part of the Paleocene. Within a few

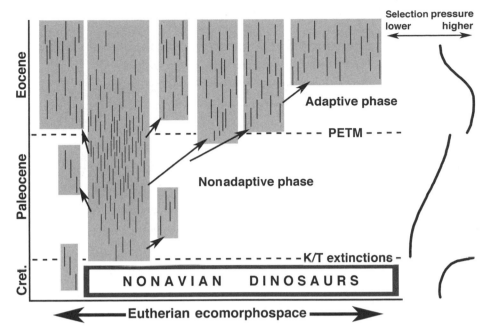

Fig. 6.6. Nonadaptive and adaptive radiations of eutherians in North America. Although there is certainly a phylogenetic overprint to the pattern shown (e.g., the appearance of modern placental orders), the arrows indicate shifts to new guilds or niches (gray rectangles) rather than specific new taxonomic groups. The short black lines signify species. The largest rectangle represents condyarths, which rapidly expanded taxonomically after the K/T boundary. This radiation was nonadaptive, or nearly so, as the condylarths showed relatively low levels of ecomorphologic differences compared to those found among modern placental orders. As natural selections intensified among condylarths, the radiation became adaptive as new guilds (and clades) began to arise, notably of modern placental orders (the rectangles in the Eocene). This was intensified and further driven by the Paleocene Eocene Thermal Maximum (PETM) and global warming.

million years or less, as many as 50 genera of condylarths and many more species are known in North America,[31] with records from Asia, Europe, and South America known not long thereafter. This agrees with John Alroy's proportional compilation, which shows that, in the earlier Paleocene, condylarths constitute over 50% of the species-level diversity in North America.

Evolutionarily, anything that survived may have thrived, radiating in a nonadaptive fashion into myriad small-to-medium-sized species varying some but not greatly from their next-closest relative. This is why doing systematic analyses of these early Tertiary taxa can be so difficult; they all look alike. For terrestrial vertebrates or at least for mammals, Darwin's natural selection was at its nadir. All this changed with the inception of the Eocene and the PETM. Natural selection began to apply its ecological screws, and within a short interval of geological time, quite new and different mammals—many of today's orders—made their debut. An adaptive radiation of mammals had begun in earnest, including going to the sea and venturing into the air for the first time as true flyers.

The mechanisms of how such a nonadaptive radiation followed by an adaptive radiation occurs are not well understood, in part because these combined events may well be unique in mammalian history. The inklings of some quite early Cenozoic mammal adaptive radiations are known; one is based on the middle Paleocene site in South America called Tiupampa. Here marvelously preserved specimens of metatherians and eutherians are found.[32] They are all of small to medium size. Unlike in North America, where eutherians in the form of condylarths probably outcompeted and led to the near de-

mise of metatherians at the end of the Cretaceous, when the metatherians reached South America they gave rise not only to modern opossum-like marsupials but also to an array of small-to-large carnivorous forms, while the eutherians gave rise to an ecological cornucopia of small-to-very-large herbivorous mammals, known as the native South American ungulates, or meridungulates. Only the opossum-like marsupials and none of the native South American ungulates survived to the present. This case of Tiupampa points up the too-often-avoided fact that single-cause hypotheses of K/T extinction paint a nearly uniform global portrait of what happened after the extinctions. The differing results of metatherian and eutherian evolution in North and South America weaken the case for a single cause.

Another case echoing that of Tiupampa is the currently earliest known site in Australia, where most metatherians—some possibly belonging to groups of living Australian marsupials—are not assignable to modern groups but rather show their South American heritage. This early Eocene site, called Tingamarra, also has bats, which shows that the adaptive radiation of placentals was well under way before the first marsupials really began their adaptive radiation in Australia.[33]

There are a number of other clearly adaptive radiations of placentals, such as monkeys or antelope in Africa, but none show the signs of an earlier nonadaptive phase, as seen in the early Cenozoic of North America. The closest we might come today for nonadaptive radiations are recent radiations of rodents and marsupials in South America, which show closely related species separated by an undulating landscape cut by Amazon tributaries.[34]

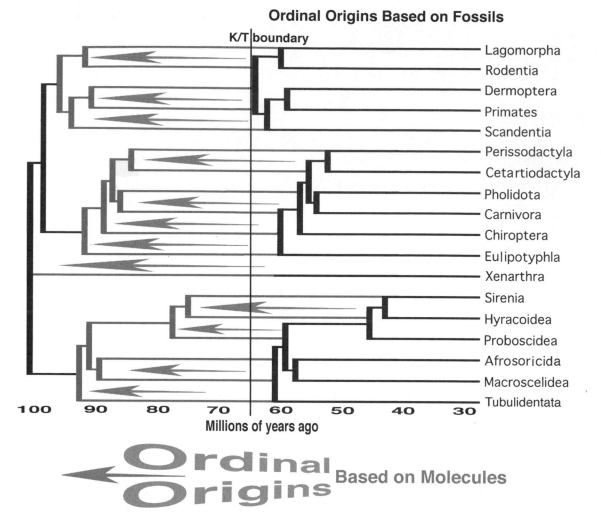

Fig. 6.7. Ordinal origins and diversifications post-K/T boundary based on the fossil record compared to those determined using molecular date.

THE ORIGIN AND RISE OF PLACENTALS— WHEN?

The fossil record unequivocally demonstrates that living orders of placental mammals did not appear or begin to radiate until after the K/T boundary, yet molecularly based studies continue to argue for a pre-K/T boundary appearance and in some cases radiation of extant placental orders. As discussed earlier, the fossil record indicates the origin of Eutheria, which is the split with Metatheria, at a minimum of between 125 and 105 million years ago, depending upon which species are accepted as eutherians. According to the fossil record, not a single order of placental mammals is known before the K/T boundary. As discussed at length in Chapter 3, attempts have been made to identify Cretaceous mammals that may have given rise to clusters of placental orders, what would be called superorders of placentals. As discussed in that chapter, we thought our zhelestids from Uzbekistan, along with zalambdalestids and cimolestids, might fit this bill. Recent phylogenetic studies point to these groups as being stem eu-

therians, not crown placentals.[35] Thus, the fossil evidence for now does not support the presence of any placentals in the Cretaceous, and most profoundly a very good fossil record shows that no placental orders existed in the Cretaceous. This is, of course, Doug Deutschman's and my Explosive Model of placental radiation.[36]

The 2007 paper by Olaf Bininda-Emonds and colleagues paints a very different portrait. All of their dates of origin for major mammalian clades are decidedly older than those based on fossils (fig. 6.7). As discussed earlier in this chapter, the fossil record is not such that it can support or refute the earlier splits within mammals argued from molecules; but the fossil record emphatically refutes the oldest molecularly based times of origin of modern mammal orders. Table 6.1 compares these considerable disparities.[37] The comparable oldest ages of orders based on fossils are liberally pushed to an older limit, and there is some dispute about what to include in any given order, but my point is not as much to show the most accurate date as to have the oldest approximation

Table 6.1. Comparison of fossil and molecularly based dates for the origin of placental mammals

Placental order	Molecularly based time of ordinal origin (mya)	Fossil-based time of ordinal origin (mya)	% greater age of molecular dates (my)	Molecularly based time of ordinal diversification (mya)	Oldest fossil reference
Rodentia	92	61	51	85	Meng and Wyss 2005
Lagomorpha	92	61	51	67	Meng and Wyss 2005
Primates	91	65	40	88	Silcox et al. 2005
Dermoptera	91	61	49	13	Silcox et al. 2005
Scandentia	94	49	92	33	Silcox et al. 2005
(Cet)Artiodactyla	87	55	58	74	Theodor, Rose, and Erfurt 2005
Perissodactyla	87	56	55	58	Hooker et al. 2005
Carnivora	85	61	39	67	Flynn and Wesley-Hunt 2005
Pholidota	85	46	85	20	Rose et al. 2005
Chiroptera	89	53	68	75	Simmons et al. 2008
Eulipotyphla	92	61	51	84	Asher 2005, McKenna and Bell 1997
Xenarthra	101	58	74	73	Rose et al. 2005
Afrosoricida	91	37	146	85	Seiffert et al. 2007
Macroscelidea	91	49	86	51	Holroyd and Mussell 2005
Tubulidentata	93	24	288	NA	Holroyd and Mussell 2005
Sirenia	76	49	55	52	Gheerbrant, Domning, and Tassy 2005
Hyracoidea	76	55	38	19	Gheerbrant, Domning, and Tassy 2005
Proboscidea	78	60	30	20	Gheerbrant 2009
			Average 57*		

*Excluding afrosoricids and tubulidentates.

that can be compared to molecularly derived dates, giving the latter dates the benefit of the doubt. All would agree that the considerably higher percentage differences between the two sorts of estimates for such orders as Afrosoricida and Tubulidentata almost certainly could be blamed on the fossil record and on the fact there is only a single species of living tubulidentate from which to extract DNA. No such agreement could be found for the discrepancies for the other 16 orders, which average a 57% difference between molecular and fossil dates of origin. What the molecular dates are arguing is not that some of the estimates of the origin of orders based on the fossil record are wrong, but that they are all wrong by an average of almost 60%!

These huge discrepancies are often attributed to the poor quality of the fossil record. It is also argued that the molecular dates show when the populations of mammals split, leading to the various modern orders, but before each evolved its ordinal characteristics, such as ever-growing rodent incisors or the slicing carnassial cheek teeth of carnivores. According to this reasoning, such ordinal features did not appear until about the time these orders show up in the fossil record. This is an interesting supposition, but it fails for three reasons. First, this reasoning is not used for just a few of the placental orders but for all 18 of them. The chance that the fossil record would be so bad for so long is vanishingly small, and on the opposite side, how could the record suddenly become so good that all but three of the 18 of the orders appear in the same 16-million-year window of time?

Second, one can point to so-called living fossils such as the horseshoe crab, the coelocanth, or the opossum, which seem to have changed little morphologically during their evolu-

tionary history. But what we are being asked to accept in this evolutionary stasis model is that *all*, not simply one or two of the 18 placental orders, existed for tens of millions of years without evolving any morphological characteristics that we use to recognize them.

Third, and most significantly, according to the molecular data the diversifications within some orders of mammals began well before the end of the Cretaceous. What this means is that, unlike the claims that most orders arose in the Cretaceous but simply did not evolve morphologically for tens of millions or years, these orders were not only around but they had also started to separate into their various lineages.

Rodentia is a good case to illustrate this incredible claim. As table 6.1 shows, the molecular data argue that rodents arose 92 million years ago. At the very least this means that the split from their nearest relatives, the lagomorphs, occurred at this time, hence the same date for the origin of both orders. The next molecular date for rodents is 85 million years ago, for the beginning of their diversification into various lineages. This would be fine, until one realizes that each of these diverging lineages would have had to possess the features that make them rodents, notably an ever-growing pair of upper and lower incisors. They would have been recognizable as rodents. This would mean squirrel-like mammals gamboling about beneath the feet of *Tyrannosaurus* and *Triceratops*. Actually, these specific dinosaurs did not appear in the fossil record for almost another 20 million years. So, according to such molecularly derived dates, full-blown rodents were on Earth before these two famous dinosaurs appeared.[38] The molecularly based dates in table 6.1 would require fully fledged members of eight other orders of placental mammals to be swimming, flying, and galloping around with dinosaurs! These claims stretch scientific credulity, conjuring Hollywood visions of cavemen battling dinosaurs.

It would be incorrect to say that all molecularly based studies place placental ordinal origination dates are as far back into the Cretaceous as does the 2007 paper by Bininda-Emonds and colleagues. Some studies place ordinal origins on both sides of the K/T boundary, and some even do not assume a constantly evolving molecular clock.[39] There are clearly issues with the older molecular dates, both in agreeing with the fossil record and with each other.

Keep in mind that ultimately all molecularly based estimates for the timing of the origin of any lineage must be tied to the fossil record. Too often the role of fossils in dating divergence times is misunderstood. Fossils can provide minimum-age estimates for any branches in a given phylogeny. They serve as constraints for establishing splits in a phylogeny.[40] It is clear from recent studies that the use of many well-established fossil dates is best, especially those occurring along a single lineage.[41] Did Bininda-Emonds and colleagues follow these rules? I would say no. Indeed, this was a huge data set of 4,500 living species, yet it used only 30 fossil dates, 27 of which belong to members of living orders. Amazingly, the base of their tree for the origin of mammals was set by a single jaw of a Jurassic-aged (166 million years ago) specimen from Madagascar named *Ambondro mahabo*. This was a rather odd species upon which to fix their tree, as its position has been all over the mammalian evolutionary map—a relative of monotremes, somewhere near the base of the metatherians and eutherians, or even within Eutheria.

It is very likely that many of the discrepancies in the timing of the origin of living placental orders, both between different molecular studies and between molecular and fossil data sets, will turn out to be a combination of three issues: fossil calibration, unwarranted assumptions about molecular clocks, and vagaries of the fossil record. I will, however, propose another contributing factor, which I feel has not been given proper attention, namely, that not only may there have been an acceleration of morphological change and speciation rates within eutherians at the K/T boundary but that this was accompanied by an unprecedented (at least for eutherian mammals) acceleration of molecular rates of evolution. Until recently the evidence for this has been circumstantial and controversial, but its does warrant consideration, and the evidence is beginning to accumulate that there are considerable differences in molecular rates of evolution.

For example, evidence has been presented for a correlation between both body size and generation time versus rates of molecular evolution.[42] Similarly, humans and elephants have large brains and longer generations compared to their nearer relatives, the mouse and the tenrec, respectively, which have much smaller brains and shorter generations. The prediction would be for similar rates of molecular evolution, at least among more closely related species, regardless of size, yet it was found that the more distantly related human and elephant lineages showed much slower rates of molecular evolution than the tenrec and mouse lineages for genes related to aerobic metabolism.[43] Another detailed study examined rates of molecular and morphologic evolution across a wide array of plant and animal taxa (dwarf dandelions, the plant *Sedum*, the birch family, beetles, dabbling ducks, caniform carnivores, salamanders, and echinoids). A considerable correlation between rates of molecular and morphologic change was found. Although a mechanism has not been clearly identified, it is possible that genetic bottlenecks (usually caused by drastic reductions in population size) may have accelerated both the molecular and the morphologic evolution.[44]

Lindell Bromham and her colleagues have examined both the general effects that such factors as body size and generation time might have on the rates of molecular evolution and the specific question of whether rates of molecular evolution speed up during major evolutionary radiations. Their findings are mixed. They argue that the rates of molecular evolution vary consistently with body size, population dynamics, lifestyle, and location of species.[45] For a much older radiation than that for placental mammals, the radiation of all animal phyla over 550 million years ago, Bromham and her colleagues found a correlation between body size and the rate of molecular evolution of mitochondrial genes. These genes occur in the mitochondria of many cells; they help control the functioning of the cell. Their findings indicate a positive effect on the rates of this molecular evolution. They concluded that if average body sizes were smaller in the early history of metazoans (animals) and that if rates of species diversification were higher, over time mitochondrial genes would have undergone a slowdown in evolutionary rate after the diversification.[46] It remains to be shown if at least some nuclear genes, which are increasingly used in molecular studies of deep relationships, also show a rate increase during species diversifications. My guess is that a relationship will be found.

One intriguing study by William Murphy and colleagues in 2003 examined rates of chromosomal changes (breaks) in placentals from the Cretaceous into the Tertiary. This study concluded that rates of chromosomal evolution within mammalian orders increased following the Cretaceous/Tertiary boundary. "Superordinal lineages predating the K-T boundary (i.e., the beginning of the Cenozoic) evolved at a rate of roughly 0.11 to 0.43 breaks per million years, whereas in ordinal and familial evolutionary lineages during the Cenozoic we find rate increases by factors of 2 to 4 in carnivores, primates, and cetartiodactyls, and by as much as a factor of 5 in rodents. . . . The only exception is the cat lineage, whose lower rate is partly a by-product of reduced map resolution relative to other species."[47] The authors' argument follows the Short Fuse Model, with very short superordinal clades followed geologically rapidly by ordinal origins within the Late Cretaceous.

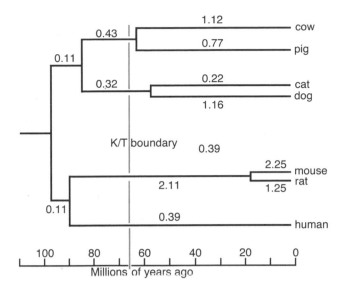

Fig. 6.8. Rates of chromosome breakage for some placental mammals (*above and below the horizontal lines*). The molecularly based times of origin follow the Short Fuse Model. See text for further discussion. Modified after Murphy et al. 2005.

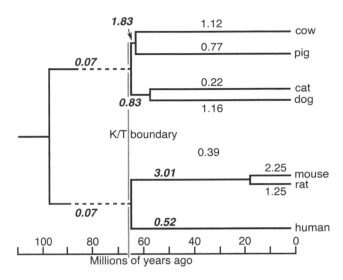

Fig. 6.9. Rates of chromosome breakage for some placental mammals (*above and below the horizontal lines*), using the post-K/T boundary origin of placental orders based on the fossil record. The new calculations, compared to those in figure 6.8, are shown italicized and bold. Because the superordinal clades originate in the Cretaceous, this is the Long Fuse Model. See text for further discussion. Modified after Murphy et al. 2005.

On closer inspection, the authors' description of their results (redrawn here in fig. 6.8) does not quite match the above description. Superordinal clades—Laurasiatheria (cow through dog) and Euarchontoglires (mouse through human)—both show 0.11 breaks per million years. The 0.43 breaks per million years to which they refer, however, are ordinal, as they occurred along the stem of Artiodactyla (their Cetartiodactyla) (see fig. 6.1). The rate of 0.32 breaks per million years shown basal to Carnivora pertains to the stems leading to Carnivora's nearest ordinal clades in figure 6.3 (Perissodactyla and Pholidota), as well as to the stem of Carnivora. Lacking further information we cannot determine which of the 0.32 breaks per million years apply to interordinal changes and which apply to (stem) ordinal changes. This uncertainty of interordinal versus ordinal rates of change also applies to Rodentia and Primates (see fig. 6.8) because we cannot determine how much of the change is interordinal, how much is ordinal, and how much is intraordinal. The split between mouse and rat is not of great help as it is a quite late intraordinal split within Rodentia. Intraordinal rates of breakage range from 0.22 for the cat up to 2.11 for murids (the mouse/rat family). As they indicate, the low value for the cat appears to be in part artifactual. When the description is corrected, providing for the limitations of the data, the authors' assertion of a difference in rates of breakage between interordinal and ordinal is, if anything, amplified further. Rates among superordinal clades (Laurasiatheria and Euarchontoglires) are only 0.11 breaks per million years, whereas in interordinal and lower levels the rates jump to a minimum of three times faster (again with the cat being less, at only twice as fast).

Interestingly, if the Explosive Model is applied, which places the timing of ordinal appearances and diversifications after

the K/T boundary (see fig. 6.9), the differences are even more profound in the rates of chromosomal breakage for superordinal taxa (Laurasiatheria and Euarchontoglires) compared to diversification of ordinal clades (Artiodactyla, Carnivora, Primates, and Rodentia). Superordinal rates are decreased by more than one-third, while ordinal rates are increased from about one-third up to a factor of almost three.[48]

Whether the Short Fuse or Explosive models of ordinal origin are employed, the message is the same: there is a profound jump in the rate of chromosomal breakages at the point after extant ordinal clades appear, if not slightly earlier. Without denser taxonomic sampling one cannot tell if the intraordinal rates in particular occurred nearer the time of the origin of the ordinal clades after the K/T boundary, later in the intraordinal radiation, or are equally parsed through the Cenozoic. One possible way of testing the thesis proposed here would be to find higher rates of change between major interordinal clades compared to lower changes for later splitting events.

While the authors of the study note that the breakpoints in the chromosomes are important in that they are reused sites in mammalian chromosomal evolution, they do not explicitly provide a reason for the jump in the rate of chromosomal evolution. It is known that chromosomal breakage can be spontaneous or induced. The former is the result of mistakes caused during DNA replication, recombination, or repair, whereas the latter is caused by various forms of radiation and various physical and chemical mutagens that raise the rate of mutations above the normal background levels. Something or things caused the increase in chromosomal breakage rates seen by William Murphy and his colleagues at the time that ordinal clades of placental mammals appeared. The most likely candidates are severe environmental perturbations occurring at the time of the K/T boundary extinctions.

If chromosomes show such an evolutionary acceleration shortly after the K/T boundary, why would not the genes that compose them? Let us start from the assumption that the molecularly derived date of around 100 million years ago for the appearances of the four superordinal clades of placental mammals—Euarchontoglires, Laurasiatheria, Xenarthra, and Afrotheria—is correct. The fossil record, although not so strong in this time interval, does suggest the origin of eutherians a minimum of between 125 and 105 million years ago, so a superordinal diversification shortly thereafter is reasonably argued. Genomic changes in each major lineage would remain relatively stochastically clocklike throughout the Late Cretaceous. At the K/T boundary and shortly thereafter the fossil record of eutherian mammals (based largely on North American records) show a five- to sixfold increase in species diversity. Although relatively few of these species ultimately survived to give rise to modern placental orders, the myriad species explosions accumulated chromosomal changes some four to five times faster than before the K/T boundary extinctions. There is every reason to believe that the rate of genomic changes also went haywire during this unprecedented interval of mammal evolution. Thus, some (most? all?) of the differences between molecularly and fossil-based times of placental ordinal appearances are in fact caused by the acceleration of genomic evolution following the K/T boundary.

There seems little doubt that bottleneck effects, when populations crash, would have been inordinately high as a result of the massive biotic reorganization across the Cretaceous/Tertiary boundary. During such tremendous episodes of biotic reorganization, both molecular and morphologic evolution might be expected to increase somewhat in concert. This hypothesis explains the paleontological pattern of extant placental orders originating and diversifying in the early Cenozoic rather than in the Late Cretaceous. Rates of morphologic and molecular change may be far more synchronized than is usually recognized, especially during times of biotic upheaval such as the K/T boundary.

We must return to the K/T boundary extinctions and remind ourselves of just how profound an effect these extinctions had in reshaping Earth's biota. The dominant terrestrial vertebrates—nonavian dinosaurs—were dead after a reign of 160 million years. The mammals, notably the eutherians (except in South America, where metatherians played a major role) began a time of explosive speciation but with only moderate levels of ecological and morphological diversification. Perhaps within about five million years selective pressures increased and we see the appearances of some modern placental orders. At the genomic level it is often argued that the accumulations of changes at this level are clocklike. Whereas the majority of molecular biologists no longer accept a universal molecular clock, it is generally accepted that there is a clocklike behavior for genes coding for some proteins. Gone are the days when it was argued that there is one molecular clock for all species, because too much evidence has accumulated that no such universal molecular clock exists. Rather, there are myriad little clocks defined within various clades. Yet the perception of some sort of overarching molecular clock persists to the point that it is hard to imagine any possible major perturbations to this system.

I think that just such a perturbation did occur with eutherian mammals at the K/T boundary, specifically, that there was what appears to have been at least a temporary acceleration of molecular clocks at this time. With the very reasonable assumption that the older fossil dates for placental ordinal origins are close to correct, I am arguing that the much older molecular origination and diversification dates for placental orders are in part caused by a temporary acceleration of molecular evolution above normal levels, averaging about 50% to 60%, during at least the first few million years when eutherians were explosively speciating. Some of these led to placentals. This scenario is supported by the changes in rates of chromosomal breakage as placental orders appeared.

THE ORIGIN AND RISE OF PLACENTALS— WHENCE?

Understanding where placentals arose and diversified has until the relatively recent past been the purview of the fossil and rock record. With the advent of radiometric dating and then the triumph of plate tectonics, our discussions of where and when various groups of animals arose became ever more accurate and precise.

The arrival of molecular systematics provided yet another powerful tool in biogeographic studies, one that has brought some real surprises for mammalian systematics and biogeography. The greatest of these was the recognition of an African radiation of placental mammals discussed earlier, which was aptly dubbed Afrotheria. The mammals that compose Afrotheria at first appear an unlikely group to be thrown together until one realizes that other superordinal groupings, such as Laurasiatheria, are composed of an equally unlikely cluster of shrews, dogs, cows, and whales. We had some inkling from the fossil record that laurasiatheres, or at least some of its members, were distant relatives, but not so for afrotheres, with one major exception. The fossil evidence had for some time grouped hyraxes, sirenians, and elephants, but aardvarks, golden moles, elephant shrews, and tenrecs? This seemed an unlikely combination if there ever was one. In addition to Afrotheria, three other superordinal groupings have consistently been found in a majority of molecular studies: Xenarthra, Euarchontoglires, and Laurasiatheria (see fig. 6.1). In turn, Euarchontoglires and Laurasiatheria are almost universally recognized as a clade Boreoeutheria, whose name, meaning "northern true beasts," indicates where it was thought to have arisen. The interrelationships of the Boreoeutheria, Xenarthra, and Afrotheria are, however, anything but settled at this time. All possible combinations have been suggested (fig. 6.10). There is hope for resolution.

Asher and his colleagues indicated that "many recent studies support Afrotheria and Xenarthra as sister taxa in Atlantogenata, but confidence in this result remains elusive. Further evaluation of the statistical support of the Atlantogenata hypothesis and scrutiny of homoplasy in an allegedly 'homoplasy-free' class of retroposon data suggest that the

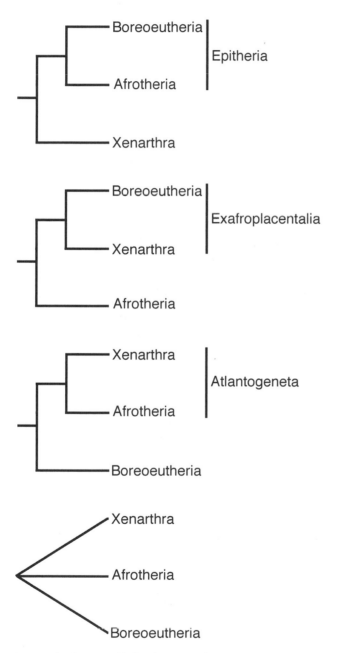

Fig. 6.10. The three possible combinations of the three superordinal clades of placental mammals for which we have strong support of monophyly. The fourth option argues that all three originated at about the same time.

placental root is still not adequately resolved."[49] *Homoplasies* are resemblances between species, ranging from the level of genes to that of organisms, which result most often from similar environments or functions, such as wings in all flying organisms. One of the "homoplasy-free" classes of retroposon data, which are argued to be unique (or nearly so) genomic changes, has, as we will see later, been used to suggest a three-way split between Boreoeutheria, Xenarthra, and Afrotheria (see fig. 6.10).

Attempts at unraveling the biogeographic patterns of these three molecularly based superordinal clades have relied almost entirely on the geological record while essentially ignoring what the fossils say. The best known of these was dubbed the "Garden of Eden" hypothesis in 2003.[50] The authors reason that at least the early stages of the placental radiation occurred in Gondwana. Gondwana in its most complete form, before it began to break apart in the mid-Jurassic some 165 million years ago, was composed of South America, Africa, Madagascar, India, Australia, New Zealand, and Antarctica. The "Garden of Eden" hypothesis suggests that Eutheria may have established a Pangaean (worldwide) distribution some 173 to 176 million years ago, after its split with Metatheria (fig. 6.11A). The splitting of Pangaea into Gondwana in the south and Laurasia (North America, Europe, and Asia) in the north some 160 to 170 million years ago left stem eutherians in Laurasia—for example, the early eutherian *Prokennalestes,* mentioned earlier in this chapter. The most recent common ancestry for the crown group Placentalia, according to this hypothesis, had its most recent common ancestry in Gondwana (fig. 6.11B). The authors argue that the earliest split, 105 million years ago, was between Afrotheria, in Africa, and the common ancestor of both Xenarthra and Boreoeutheria, in South America. The primary line of evidence is the agreement of the authors' molecular clock of 105 million years ago for the above split with the estimated sundering of Africa and South America between 100 and 120 million years ago (fig. 6.11C). Finally, there was a dispersal of ancestral boreoeutheres to North America after its ancestor had split from xenarthrans, between 88 and 100 million years ago (fig. 6.11D).

Paleomammalogists John Hunter and Christine Janis have delineated the problems with Gondwanan or southern origins scenarios.[51] They first note that one cannot reconstruct the place of origin of Placentalia using phylogenetic reasoning unless outgroups are incorporated into the analysis, which in this case would include earlier occurring eutherians basal to placental mammals. When this is done, a Laurasian, or northern, origin, not a Gondwanan, or southern, origin requires the fewest number of steps and assumptions. They next point out that a purely vicariant, or splitting, mode for the earliest placental mammals requires that the earliest divisions within Placentalia happened much earlier, in the Late Triassic and Early Jurassic, when all the continents were still together as Pangaea and well before even the oldest molec-

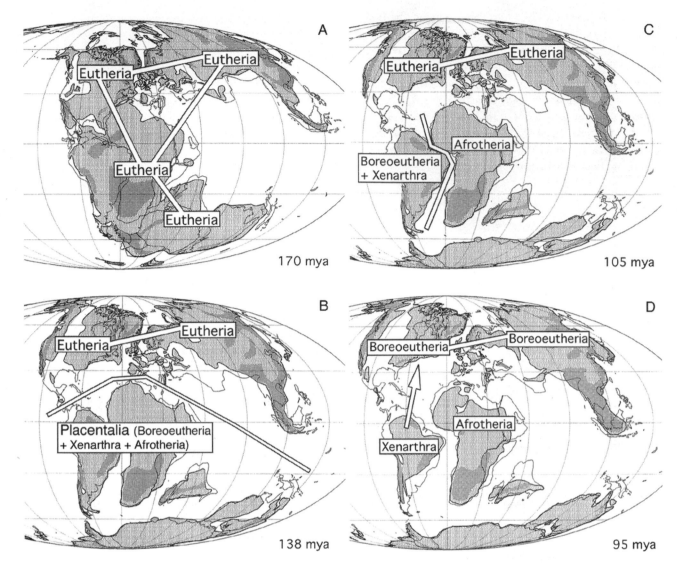

Fig. 6.11. The "Garden of Eden" hypothesis for the biogeographic history of the three superordinal clades of placental mammals. See text for explanation. Modified after Archibald 2003.

ular divergence estimates for Placentalia. This is at a time closer to the origin of mammaliaforms and Mammalia. Finally, they note that Gondwanan-origin models would require at least twice the number of dispersal events compared to Laurasian-origin models.

In a recent molecular study Hidenori Nishihara and his colleagues claim that a near-simultaneous appearance of the three major superorders of placentals—Afrotheria, Xenarthra, and Boreoeutheria (see fig. 6.10)—fits with the similarly timed breakup of Africa, South America, and Laurasia.[52] Using what they argued to be rare genomic changes called *retroposons*, the authors found that all possible splits of Afrotheria, Xenarthra, and Boreoeutheria were almost equally likely, leading them to conclude that the split between the three groups occurred at almost the same time. Turning to geology, they claimed that new evidence showed that Africa and South America were last connected 120 million years ago

through the Brazilian Bridge and that Laurasia and Africa might have been connected through Gibraltar until 120 million years ago (fig. 6.12). Again, fossil evidence was not used.

The authors seemed particularly enamored of the idea that continents must be in contact for dispersals to occur, followed by vicariance (splitting) events once continents had separated. There are simply too many instances from the fossil record to show that this is not the case, but unfortunately, once again the mammalian fossil record was ignored. Two related cases that readily come to mind are South American monkeys (platyrrhines) and the great diversity of South American rodents (caviomorphs), which include such well-known rodents as guinea pigs and chinchillas. Fossils indicate that the ancestors of both probably reached South America at least 50 million years ago, yet there were no land connections at that time with either North America or Africa, the two most probable sources. Anatomical studies of fossils from Africa point to this area as the source for both groups, but the migration to South America must have transpired over open water.

If fossils are included, interesting biogeographic patterns emerge. Both branches of Theria—Metatheria and Euthe-

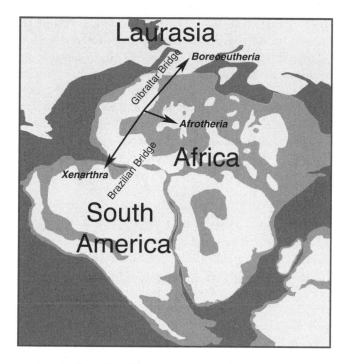

Fig. 6.12. An illustration of the hypothesis that the three superordinal clades arose and reached Laurasia, Africa, and South America at nearly the same time. White is land, darker gray is deep ocean, and lighter gray is shallow ocean. Modified after Nishihara, Maruyamab, and Okada 2009.

ria—show similar biogeography until the early Cenozoic, about 55 million years ago, when marsupials begin their diversification in Australia unhindered by placentals, with the notable exception of bats (fig. 6.13). Placentals dominate the remainder of the world, except where marsupials and placentals share South America, from the early Cenozoic onward. Both the earliest metatherians and eutherians are known from Laurasia by at least 100 and possibly 125 million years ago. Yet, no metatherians or eutherians have been confidently reported from any Gondwanan continent before the K/T boundary except for a single partial lower molar from the Late Cretaceous of Madagascar,[53] and a small clade of eutherians from the Indian continent,[54] which may have begun colliding with Asia by this time.

Granted, the mammalian fossil record from the Cretaceous of much of Gondwana remains woefully inadequate compared to that of Laurasia. For South America, however, and less so for Australia, we do have Cretaceous mammals. *None* appears to be metatherian or eutherian. Because their stem lineages are not present before the K/T boundary in Gondwana, the same holds for marsupials and placentals—somewhat like the old joke that if your parents do not have children, neither will you. Both metatherians and eutherians do start showing up on all former Gondwanan continents after the K/T boundary, with marsupials and placentals by some 60-plus million years ago, if not earlier, as suggested by the mid-Paleocene site of Tiupampa, in South America. As with the issue of the timing of the origin and diversification of placentals, ignoring the fossil record is simply ignorant. Importantly, the message from the fossil record is the same for the biogeography of placental mammals, or at least extant orders, and it is one of a largely post-K/T boundary diversi-

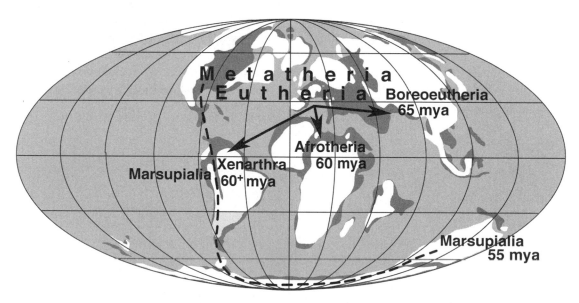

Fig. 6.13. Biogeographic history of Metatheria (including Marsupialia) and Eutheria (including Xenarthra, Afrotheria, and Boreoeutheria) based on the fossil record. Dates are earliest records. The earliest records for both Metatheria and Eutheria are between 100 and 125 million years ago in Laurasia. White is land, darker gray is shallow ocean, and lighter gray is deep ocean. Base map originally from Smith, Smith, and Funnell 1994.

fication from Laurasia. Marsupials are less well known, but where known, they always agree with the eutherian record, with one major exception. Why did placentals (not counting bats) not reach Australia with marsupial ancestors some 60 million years ago? Even if we find their fossils in Australia at this early stage, it would be difficult to determine why marsupials survived but placentals did not.[55]

CODA

There is nothing to tie nonavian dinosaur extinction to any aspect of mammals or their radiation, yet mammals are undoubtedly the greatest evolutionary beneficiaries of the extinction of nonavian dinosaurs. Few other episodes in the history of life have rivaled the dramatic extinction of nonavian dinosaurs and the rapid radiation of mammals. Unquestionably, without the former's extinction, the latter's diversification into clades as different as bats and whales would never have occurred. While living reptiles, including birds, outnumber living mammals by more than three to one, their range in size does not come close to rivaling that of mammals. Even if nonavian dinosaurs are added, mammals still hold the record. Nothing rivals the size of a living blue whale, unless possibly an extinct blue whale relative. Birds and mammals have mastered endothermy, and some other reptiles and most mammals give live birth, but in the case of reptiles, with little support from maternal tissues. No female reptile (again including birds) can brave Arctic waters while nurturing a maturing embryo in her body. In a combination of reproduction and physiology, mammals are a great evolutionary and ecologic success story.

Despite what one reads, sees, or hears in the media, we are only now beginning to have some true inkling of why and how rapidly the nonavian dinosaurs disappeared and when and how the mammals radiated. What we can say with considerable confidence based on the North American fossil record is that nonavian dinosaurs decreased in ecological diversity and taxonomic diversity by at least 40% in the last 10 million years of the Cretaceous. How these declines played out in the probably total extinction of nonavian dinosaurs by the end of Cretaceous is not certain. One can surmise that

the receding global seaways and increased volcanic activity helped place a strain on the already decreased nonavian dinosaur diversity, with an extraterrestrial impact being the final and in some cases violent cause of the extinction of the last nonavian dinosaurs. It is known that the decrease in seaways caused a concomitant decrease in coastal plain habitats, from which all latest Cretaceous nonavian dinosaurs are known, while increased volcanic activity is almost certainly the major cause of latest Cretaceous cooling. The physical and biological manifestations of an extraterrestrial impact have proved harder to model and test. Acid rains and global wildfires were popular scenarios that have been shown to be almost certainly wrong. Extended darkness caused by material suspended in the atmosphere after the impact remains the likeliest culprit for the devastating effects on plant and animal life, at least in some areas. A burst of infrared radiation following the impact may have also played a role in some regions, but more recent modeling shows that it was not the global catastrophe once envisioned by its authors.

Whatever the causes of these extinctions were, the major result on the land for vertebrates was the replacement of nonavian dinosaurs by mammals. On land, mammals never rivaled nonavian dinosaurs in size, but within five or so million years mammals underwent an explosive radiation in numbers of species if not in ecological diversity. By 10 million years after nonavian dinosaur extinction, however, mammals underwent an even more impressive ecological radiation, which gave us all our major clades of modern mammals. According to molecular results using molecular-clock models, the latter radiation began tens of millions of years before the extinction of nonavian dinosaurs. The fossil record, however, demonstrates that this simply was not possible, unless we have missed the squirrels cavorting with dinosaurs. It is far likelier that the dramatic events leading to the demise of nonavian dinosaurs also greatly perturbed the molecular clocks of the surviving mammal lines, causing them to speed up as the mammals explosively radiated. For the past 65 million years mammals have continued an evolutionary odyssey set in motion by the extinction of their ecological nonavian dinosaur precursors.

Epilogue
Lessons from the Past

THE EVENTS DESCRIBED in this book and the hypotheses that have been developed to explain them appear so remote in time that it is hard to see how they have any relevance to here and now. This is compounded by the reality that each human life is so short that in the scheme of things we cannot conceive of how we fit into or affect most of anything around us, yet we have the hubris to believe we might be able to keep another object from space from smashing into Earth. We actually do have the wherewithal to change the course of life on this planet. We have been doing it for millennia but have only realized this power very recently. Apart from calamities, such as extraterrestrial impacts, earthquakes, volcanic eruptions, and storms about which we have little or no understanding of how to control, we understand all too well that many, if not all, of the ills we have been inflicting upon ourselves and this planet stem from one root cause: overpopulation. A brief parable illustrates this far better than reeling off figures.

AN IMAGINARY TALE

According to the Kurdish historian Ibn Khallikān, the game of chess was invented by the Indian Sissa Ibn Dàhir to amuse Ardashīr-i Pāpagān (third century CE), the king and founder of the Sassanid Empire, the last pre-Islamic Persian dynasty. Ardashir was so pleased with chess that he ordered the game to be placed in the temples, considering it the best method to introduce the art of war. Wishing to show his gratitude, he asked Sissa to request whatever he desired. Sissa replied that he wanted a grain of wheat to be placed on the first square of the chessboard, two in the second, and that the number of grains be progressively doubled till the last square has been attained. The king considered such recompense too little, reproaching Sissa for asking for so inadequate a reward. Sissa persisted in his demand, with the king finally agreeing to grant his request. When the king's clerks learned of the request they declared to him that all the wheat in the world would be insufficient to make up that quantity he had requested. Quite angered, he ordered them to prove what they said by a series of multiplications and calculations—one grain is placed on the first square, two on the second, four on the third, eight on the fourth, and so on until the last square is reached. By the sixteenth square the number 32,768 has been reached. By the 32nd square more than four billion grains of wheat would be required, or about 100,000 kilos of wheat. Now Sissa did not seem so stupid. Even so, King Ardashir was willing to pay up, but he soon learned how clever Sissa actually was: in order to complete the 64 squares of the chessboard, one would require wheat totaling six times the weight of all the living things on Earth.[1]

A REAL STORY AND ITS CONSEQUENCES

While the doubling in Ibn Khallikān story may seem beyond comprehension, there are real patterns that follow a similar, although not identical, pathway. One of these is human population growth. Estimates vary, but in 1650 the global population may have been about 500 million people. By 1850 (200 years later), it was one billion. Eighty years later, in 1930, it had doubled again, to two billion, whereas only 50 years later, in 1980, it had more than doubled, to four and one-half billion. By the beginning of the new millennium, six billion people occupied the planet, reaching about 6.8 billion by mid-2009.[2] Estimates again vary, but by 2050 there could be anywhere from nine to 12 billion people. Even if only the lower number is reached, as now seems likely, it may not be a pretty sight. There will be many consequences of this increase in human population, but in keeping with the theme of this book, I will end by reviewing some of the findings of what is thus far the most comprehensive study of the state and fate of mammals.

In 2008, Jan Schipper and more than 100 colleagues published an article, "The Status of the World's Land and Marine Mammals: Diversity, Threat, and Knowledge." The only good news is that the diversity of mammals is richer than previously thought, with 349 species of mammal discovered since 1992. The understated title did not indicate the rather dire situations that mammals face in the coming years, but the data and words of the authors did just this. An accompanying commentary summarized how bad things are. There are 188 critically endangered species of mammal, and 29 of them, such as the baiji, a Chinese freshwater dolphin, are almost certainly extinct. The plight of large land mammals, particularly hoofed animals and primates in South and Southeast Asia, is worse. Marine mammals are at particular risk in the North Pacific, the North Atlantic, and in Southeast Asia. Extinction is likely for 1,139 of 4,651 better-known species, and another 836 species are simply too poorly studied to determine conservation needs. About half the number of mammalian species is declining in numbers, including about one in five that are not at risk of extinction now.[3] "Our results paint a bleak picture of the global status of mammals worldwide. We estimate that one in four species is threatened with extinction and that the population of one in two is declining."[4]

If one in four species of mammal is threatened with extinction and the population of one in two is declining today, with 6.8 billion people on the Earth, what will the pattern look like by 2050, with a minimum of nine billion people? It is frightening but fairly safe to say that we may witness 50% or more of mammalian species disappearing in the next 100 years, even if the human population reaches only nine billion by 2050. For mammals this would be a mass extinction occurring in the wink of a geologic eye. We will certainly have a biologically impoverished world and our very existence could be threatened. The signs are not good that we will be doing anything about this situation in the near future. We seem hell-bent on remaining passive players, waiting for the Earth to show us what it has in store for us.

NOTES

Preface

1. Jones 2008.
2. Archibald 1996.
3. Archibald 2006, 21.
4. Erwin 2006, 58.
5. Erwin 2006, 17–18.
6. Any date for the fall of the Roman Empire is of course arbitrary, but a commonly used date is 476 CE for the fall of the Western Roman Empire, while the Eastern Roman Empire did not fall until the Ottoman Empire captured Constantinople in 1453.

Chapter 1: The Late Cretaceous Nonavian Dinosaur Record

1. There is some controversy as to whether to use *Paleogene* and *Neogene* for the major divisions of the Cenozoic Era rather or the older terms *Tertiary* and *Quaternary*. Because *Cretaceous/Tertiary*, or *K/T*, boundary is so engrained in the literature, it makes no sense to shift to *Cretaceous/Paleogene*, or *K/P*, boundary. See Salvador 2006, 27.
2. As I discuss more fully in Chapter 2, and as is widely accepted, Dinosauria includes Aves, or birds. I was talked out of using the phrase *nonavian dinosaur* in a 1996 book I did on dinosaur extinction (Archibald 1996) because it was thought too cumbersome. I regret this. A slightly more cumbersome but accurate term is better than a shorter but inaccurate term. In this book, when I write *dinosaurs* or *Dinosauria*, I include birds, but *nonavian dinosaurs* excludes birds.
3. Two words that I have used as antonyms in the book are *extinct* and *extant*. *Extinct* is understood by most people, but just to be sure, I mean by it that all members of a species, genus, family, etc., are thought to no longer be in existence—all are dead. *Extant* means basically the opposite, that living members of a species, genus, family, etc., are thought to exist. I prefer this to the words *living* or *modern*.
4. Zhonghe, Barrett, and Hilton 2003.
5. Buffetaut 2003, 183. Kevin Padian has commented on this issue in his review of Buffetaut's book (2004, 13):

   > Perhaps the most astonishing aspect of Buffetaut's book is how little attention he gives to the many studies of the pace of vertebrate change across the K/T boundary in North America (except that of Fastovsky and Sheehan, which counters all the others and with which he obviously agrees). The extensive and eminently fair explanations of all the evidence by David Archibald in his 1993 [*sic*] book *Dinosaur Extinction and the End of an Era* are not even considered, just dismissed as attributing everything to sea-level regressions. None of the actual work by Archibald, W. A. Clemens, L. Bryant, J. H. Hutchison, L. Dingus, D. Lofgren, and others, which meticulously showed the pace of change of all vertebrates down to the genus level across the K/T boundary in Montana, is even mentioned. Buffetaut seems to think that such chauvinistic US-based studies are myopic and can be disregarded as not representative of worldwide change (as if anyone ever said they were). On the other hand, the other parts of the globe that he mentions, from France to India, that do preserve Late Cretaceous sediments do not have them in such abundance; nor do they have anywhere near the stratigraphic control as the North American sections; nor have they been anywhere near as extensively studied.

6. Buffetaut 2003, 183.
7. A very nicely produced volume published in 2009 titled *Mesozoic Terrestrial Ecosystems in Eastern Spain* (Alcalá and Royo-Torres 2009) shows that this region holds promise for fossiliferous sections across the K/T boundary.
8. Oms et al. 2007, 45.
9. Currie and Koppelhus (2005) provided up-to-date data on geology (Eberth), ornithischians (Ryan and Evans), and saurischians (Currie) of Dinosaur Provincial Park, Alberta. This section draws heavily from chapters in this book as well as tabulations from Weishampel et al. 2004. Where there are differences between sources, these are noted.
10. For many of us, the Dinosaur Park Formation was known as the Judith River Formation. Eberth (2005, 54–56) details the history of how this name change occurred.
11. Sources vary as to the origin of Campanian, but Alexander Averianov assures me that the source is not the village Champagne in the department of Charente-Maritime, as is often cited; rather, it was named by H. Coquand in 1857 after hills in the *"Grande Champagne,"* and the type locality is near the castle Aubeterre-sur-Dronne, which is in the neighboring department of Charente.
12. In 2009 Longrich and Currie named and described *Hesperonychus elizabethae* from the 75-million-year old Dinosaur Park Formation. At 50 cm, it is North America's smallest known theropod dinosaur.

13. Currie 2005, 370.

14. Kuiper et al. 2008, 500.

15. Based on Cifelli et al. (2004, 26), the Lancian could be as much as three million years in duration. Wilson (2005, 57) "estimated depositional duration of 2.1 Ma [million years] for the Hell Creek Formation." If this represents most of the Lancian, then the duration of the Lancian would be less than is indicated by the Lance Formation.

16. Fastovsky et al. 2004, e75.

17. Fastovsky et al. 2004, 877; Wang and Dodson 2006, 13601; Lloyd et al. 2008, 2483.

18. Barrett, McGowan, and Page 2009, 2667.

19. Barrett, McGowan, and Page 2009, 2671.

20. Fastovsky et al. 2004, e75.

21. Eberth 2005, 58.

22. These formations were identified mostly as Oldman Formation by Geological Survey of Canada (1965a,b); Jackson (1981), Geological Highway Map of Alberta, identified these formations as Oldman Formation in the south and Belly River Group in the north; Rogers (1998, fig. 1) identified them as Judith River and Two Medicine formations. Eberth (2005, 54–56 and fig. 3.2) indicates the equivalence of the Dinosaur Park Formation with much of the Oldman Formation and part of the Judith River as used in earlier sources. Areas determined from digitized version using ImageJ (2008, v. 1.40g).

23. Hartman (2002, fig. 2) identified as Hell Creek and Lance formations. Area determined from digitized version using ImageJ 2008, v. 1.40g.

24. Fastovsky et al. 2004, e75.

25. Archibald 1996, 147–64.

26. Haltennorth and Diller 1992, based on a count of species listed in this book.

27. Whitaker 1980, based on a count of species listed in this book.

28. Kingdon 1982, 530.

29. Hornaday 1886. Under the subheading of "Abundance" Hornaday wrote, "Of all the quadrupeds that have lived upon the earth, probably no other species has ever marshaled such innumerable hosts as those of the American bison. It would have been as easy to count or to estimate the number of leaves in a forest as to calculate the number of buffaloes living at any given time during the history of the species previous to 1870. Even in South Central Africa, which has always been exceedingly prolific in great herds of game, it is probable that all its quadrupeds taken together on an equal area would never have more than equaled the total number of buffalo in this country forty years ago" (387). His estimate of four million in this one herd comes from the following passage, "If the advancing multitude had been at all points 50 miles in length (as it was known to have been in some places at least) by 25 miles in width, and still averaged fifteen head to the acre of ground, it would have contained the enormous number of 12,000,000 head. But, judging from the general principles governing such migrations, it is almost certain that the moving mass advanced in the shape of a wedge, which would make it necessary to deduct about two-thirds from the grand total, which would leave 4,000,000 as our estimate of the actual number of buffaloes in this great herd, which I believe is more likely to be below the truth than above it" (391).

30. Whitaker 1980, 667.

31. While larger theropods were certainly carnivorous, theropods such as ornithomimids were omnivorous or herbivorous, and oviraptorosaurs such as *Chirostenotes* may have been insectivorous.

32. Barrett et al. 2009, online.

33. Fastovsky et al. 2004, e75.

34. Archibald and MacLeod 2004, 4 and table 1. The basis for these occurrences came from Weishampel et al. 2004.

35. Hurlbert and Archibald (1995) demonstrated that the analysis of Sheehan and colleagues (1991) was at taxonomic and sampling levels too crude to establish whether nonavian dinosaurs declined gradually or underwent sudden extinction. Even more troubling is that most specimens were not collected; thus their empirical findings cannot be verified. This problem was perpetuated in a Sheehan et al. study in 2000. As Pearson et al. (2001, 42) note about this study, "Unfortunately, they did not collect voucher specimens and have not published precise locality and stratigraphic data."

36. Pearson et al. 2001, 40; Pearson et al. 2002, 145–51.

37. Pearson et al. 2001, 42.

38. Haile et al. 2009.

39. Pearson et al. 2002, esp. 158 and figs. 4 and 5.

40. Pearson et al. 2002, 145.

41. Bailey et al. 2005, 386.

42. This and the subsequent quotes were from Wilson's Web page on the K/T boundary—http://protist.biology.washington.edu/GPWilson/Mammalian_change.htm—but much of the information is also found in Hutchison et al. 2004 and Wilson 2005.

43. Horner 2009.

44. See note 42.

45. Archibald and MacLeod 2007, 7–8 and fig. 2, after Hutchison et al. 2004.

Chapter 2: In the Shadow of Nonavian Dinosaurs

1. Hu et al. 2005, 149.

2. Rowe 1988, 247–50.

3. For birds, 10,000 species are reported by LePage 2004; for mammals, 5,416 species are reported by Wilson and Reeder 2005, xxv.

4. See note 3.

5. A short history of these events is provided by Kielan-Jaworowska, Cifelli, and Luo 2004, 6.

6. Rowe 1999, 283.

7. A detailed account of the early discovery and interpretation of Mesozoic mammals can be found in Desmond 1984, 7–16.

8. Torrens 1992, 40–44.

9. Owen 1842, 103.

10. McGowan 2001, 178–80.

11. Kielan-Jaworowska, Cifelli, and Luo 2004, 216; Rougier et al. 2007.

12. The concept of uniformitarianism held by Lyell had two parts. Processes operating today operated in the past; and changes occurred in a generally slow, stately manner. Today, we accept only the first precept.

13. According to Lyell (1872, 161–62), "We should then have monotremata in the Primary [Paleozoic], marsupials in the Secondary [Mesozoic], and placentals in the Tertiary strata, assuming for the present that the class to which *Stereognathus* belongs is still undetermined." In fairness to Lyell, by the date of this quote, he was, under Darwin's influence, beginning to accept that there was progression in the fossil record. Unfortunately, his newer views were clearly advocating a *scala naturae*.

14. Desmond (1982, 193–201) describes at some length Huxley's ideas on Paleozoic mammals and speculates as to why he rejected Owen's mammal-like reptiles as intermediates between amphibians and mammals. In addition to the general animosity

between the two, Owen was tightfisted with his fossils and thus would have been especially reticent to allow Huxley open access, while for his part, Huxley would not have liked to credit Owen as having intermediary forms.

15. Kielan-Jaworowska, Cifelli, and Luo 2004; McKenna and Bell 1997.

16. Crompton et al. 2008.

17. Preston 1993, 97–98.

18. Mayor 2000, 48–51.

19. Gregory and Simpson 1926, 14.

20. Clemens 1964, 1966, 1973; Lillegraven, 1969. Fox wrote a series of papers concerned with Cretaceous Canadian mammals before 1980 (see References). In addition, all three of these individuals have contributed substantially to this field since 1980.

21. Nessov (1997) references all earlier papers dealing with mammals from Middle Asia (present countries of Kyrgyzstan, Tajikistan, Turkmenistan, Uzbekistan, and southern Kazakhstan).

22. Lillegraven, Kielan-Jaworowska, and Clemens 1979.

23. Kielan-Jaworowska, Cifelli, and Luo 2004, 6–7.

24. Wible, Novacek, and Rougier 2004.

25. Wible et al. 2009. Novacek et al. (1997) described a not-too-distant relative of Maelestes, Ukhaatherium, in a shorter paper in Nature. It is very similar and closely related to the previously described Asioryctes. Interestingly, it preserves epipubic bones, a pair of small bones at the anterior margin of the pelvis that are typically associated with marsupials but are found much deeper in mammalian phylogeny. All living eutherian mammals (the placentals) lost these bones.

26. Archibald and Averianov 2001, 2005, 2006; Archibald, Averianov, and Ekdale 2001; Averianov and Archibald 2005; Nessov, Archibald, and Kielan-Jaworowska 1998.

27. Godinot and Prasad 1994; Prasad et al. 1994; Prasad and Sahni 1988; Prasad et al. 2007.

28. Kusuhashi, Ikegami, and Matsuoka 2008; Setoguchi et al. 1999.

29. Gheerbrant and Astibia 1994, 1999.

30. Luo 2007.

31. Ji et al. 2002 (Eomaia) and Luo et al. 2003 (Sinodelphys).

32. Cifelli 1999. Although quite possibly a eutherian, questions as to the affinities of Montanalestes still remain.

33. Fox 1984 and 1978, Johnson and Fox 1984, and Storer 1991 (Canada); Archibald 1982, Hunter and Archibald 2002, Lofgren 1995, and Montellano 1992 (Montana); Hunter and Archibald 2002 and Hunter and Pearson 1996 (Dakotas); Lillegraven and Eberle 1999 and Lillegraven and McKenna 1986 (Wyoming); Cifelli 1990 (Utah); Flynn 1986 and Rigby and Wolberg 1987 (New Mexico); and Cifelli 1994 and Rowe et al 1992 (Texas).

34. Clemens 1980.

Chapter 3: In Search of Our Most Ancient Eutherian Ancestors

1. As explained in Chapter 2, a crown clade is the living part of a much larger and often mostly extinct clade. For example, Aves is the crown clade of Dinosauria. Among mammals, Monotremata is the crown clade of Prototheria, Marsupialia is the crown clade of Metatheria, and Placentalia is the crown clade of Eutheria.

2. Wilson and Reeder 2005, gleaned from table 1.

3. Fox 1989.

4. Cope 1881, 1018.

5. Examples of shared derived characters that are used to unite all mammals are hair and the presence of mammary glands.

6. URBAC stands for Uzbekistan, Russia, Britain, America, and Canada.

7. Chester et al. 2007, 2008; Sargis et al. 2009.

8. Gheerbrant and Astibia, 1994, 1999.

9. Cifelli 1990.

10. Averianov, Archibald, and Martin 2003.

11. Setoguchi et al. 1999; Kusuhashi, Ikegami, and Matsuoka 2008.

12. Pereira et al. (2006) reviewed the natural selection and molecular evolution in primate PAX9 gene and its relation to such events as the agenesis of the third molar. "Because of the dramatic lifestyle and diet shift experienced since the discovery of fire and the development of cooking utensils, third molars, which could have been essential for the survival of earlier hominids, became not only functionless but also an important cause of morbidity for modern humans" (5680).

13. As indicated by Novacek et al. (1997), the 75-million-year-old Asioryctes and Ukhaatherium have five upper incisors as well as four lower incisors. A mammal identified in 2002 as the earliest eutherian, Eomaia ("dawn mother"), from China some 125 million year ago, is alleged to have five upper incisors (Ji et al. 2002). Except for a few photographs of this very flattened specimen, however, the authors have provided only interpretative illustrations, and notably, there are no illustrations of the upper dentition. Until such time as the specimen is properly illustrated, we must remain skeptical as to the identification of anatomical structures, especially the dentition of the animal, which is a key component of its identification as a eutherian.

14. Scott and Turner (2007, 33–34) indicate that "in nonhuman primates, lower canines typically have two roots. This condition is rare or absent in human populations, but it does attain a frequency or 5–10% in some groups." They also illustrate a human two-rooted lower canine (fig. 2.13). I doubt the commonness of two-rooted canines in nonhuman primates indicated by these authors. Swindler (2002, 20) provides a number of references indicating that two-rooted canines are uncommon in primates, but he does note one case of a frequency of 40% of two-rooted maxillary canines in Macaca fuscata (an Old World monkey) females, but the complete absence of this trait in males in this population. Because of sexual dimorphism, the males most certainly would have larger canines. Interestingly, it is among smaller zhelestids that we find two-rooted canines. Among living placental mammals, dermopterans, the so-called flying lemurs, which neither fly (they glide) nor are lemurs, have two-rooted canines as the normal condition, and along with tupaiids, or tree shrews, are the nearest relatives of primates.

15. Matt. 16:18 (King James Version).

16. Ekdale 2008, 74A.

17. Szalay and Sargis 2006, 207; Chester et al. 2007, 58A; Chester et al. 2008, 53A.

18. Luo et al. 2003, 1938.

19. Wible et al. 2007, 1004–5. A quite unusual result of this analysis is the clustering of the primate Purgatorius and the condylarths Protungulatum and Oxyprimus just outside of Placentalia, which does include other primates and condylarths. The authors note that, "although branch support for many basal nodes is weak, we examined various competing hypotheses of association using a Wilcoxon rank sum (also known as Templeton) test, and found all (except Purgatorius with Primates, and Palaeocene and Eocene 'condylarths' with Cetartiodactyla) to be significantly rejected using our morphological data set (see Supplementary Information) (1004)." More recently Spaulding et al. (2009) found that Protungulatum is one of four condylarths that is sister to a clade including Perissodactyla (odd-toed ungulates,

such as horses and rhinos) and Artiodactyla (even-toed ungulates, such as deer and cows, as well as whales). In turn, the next-closest extant sister clade is Carnivora; thus, according to this analysis, *Protungulatum* is a member of the crown clade Placentalia in a position basal to an ungulate clade of artiodactyls and perissodactyls.

20. Archibald, Averianov, and Ekdale 2001, 63–64.
21. Asher et al. 2005; Meng and Wyss 2001; Meng, Hu, and Li 2003.
22. Kielan-Jaworowska 1978, 29–32.
23. Gregory and Simpson 1926, 1–3. While these authors used *placental* in a broader sense than is now used by most, their wording suggests that they probably thought that zalambdalestids were related to living placentals.
24. Simpson (1928a, 11) made the placental tie clear when he wrote that zalambdalestids represent "a very early and non-ancestral branch" of erinaceomorphs—the clade that includes hedgehogs. This would place zalambdalestids within Placentalia.
25. Van Valen 1964, 488–90.
26. McKenna 1975, 33.
27. Archibald, Averianov, and Ekdale 2001, 63.
28. Wible et al. 2007, 1005.
29. Wible et al. 2007, 1005.
30. Archibald and Averianov 2006.
31. Lillegraven 1969, 86–87.
32. Wilson and Reeder (2005) recognized 29 orders of living mammals. With one exception I agree with their assessment. The exception is Cetacea, which must be placed within Artiodactyla. I have no objection if this is called Cetartiodactyla to reflect this change.
33. Fox 1979, 119–23; 1984, 9–14.
34. Nessov 1993, 105–33.
35. Archibald and Averianov 2001, 548–49.
36. Fox 1979, 123.
37. Wible et al. 2007, 1005.

Chapter 4: Patterns of Extinction at the K/T Boundary

1. Multituberculates were a highly successful clade of mammaliaforms that are placed inside Mammalia, closer to therians than to montremes if one accepts postrcranial evidence, outside of Mammalia if one prefers dental and cranial data, or closer to monotremes than to therians again if one prefers cranial evidence. Wherever their true relationships lay, they are known from at least the Jurassic, some 160 million years ago, until the latest Eocene, some 35 million years ago. First rodentlike early primates and then true rodents were probably their undoing. Multituberculates are by far best known from Laurasia, but with the report in 2009 (Rich et al.) of a multituberculate from the Early Cretaceous of Australia along with meager finds in South America, Africa, and Madagascar, the global occurrence of multituberculates is most likely.
2. Archibald 1996, table 5.1.
3. Archibald 1996, table 5.1.
4. Fox and Naylor 2003.
5. Clemens, pers. comm., and Zhang 2009, respectively.
6. Spaulding et al., 2009; O'Leary, 2010.
7. Wilson (2005) has a slightly different take on mammal survival at the K/T boundary, so I quote him here. The following is from his 2005 paper (66–67) detailing some of the concerns of how to count mammalian lineages approaching, at, and after the K/T boundary. Our interpretations are quite similar other than I tend

to count the survival of a clade as a K/T survival, what he terms *pseudoextinctions*.

> Disappearance events occur infrequently through the Hell Creek Formation. . . . *Parectypodus foxi* and *Cimolomys ?trochuus* disappear before the upper part of the formation, while 27 of the remaining 28 species disappear in the upper 5 m of the formation (i.e., last ~115 ky of the Cretaceous). More precisely, 11 disappear in the 82–88-m interval and 16 disappear in the 88–93-m interval, a pattern which is consistent with a stepwise extinction or extirpation or both prior to the K-T boundary and correlated with the rapid cooling from 65.7 to 65.58 Ma. However, caution must accompany interpretations, as pseudo-extinction artificially inflates the number of disappearances in the upper 5 m of the formation. . . . At least five Cretaceous lineages that seem to disappear actually persist into the Paleocene of Garfield County as descendant species. . . . Conversely, low sampling intensity in the upper 5 m of the formation (36 specimens) suggests that edge effects or the Signor-Lipps effect may mask a pattern of disappearance that is concentrated even higher in the section and indicative of a sudden not gradual extinction.

8. Archibald 1996, table 5.1.
9. Brochu 1997, 2003.
10. Feduccia 1996; Chiappe and Witmer 2002; Padian 2004; Chiappe 2007.
11. Feduccia 1995 and 1996.
12. Walker 1981.
13. Padian 2004, 212–14.
14. I added to this diversity of enantiornithines before Walker named the group in 1981. In 1975, while still a PhD student at UC Berkeley, I found parts of a nearly complete tarsometatarsus, the fused ankle and foot bones found in birds and a few nonavian dinosaurs. As in most theropods, including birds, there were three digits evident, which were fused in the upper part. When I showed the bone to ornithologists they said it was from a nonavian dinosaur because, unlike birds, it was fused at the top, not the bottom. When I showed it to dinosaurologists they said it was a bird because of the fusion. It was eventually named *Avisaurus archibaldi* and placed within Theropoda. Later it was assigned to this new clade of enantiornithines, which in the broader sense are, like all birds, theropods. I had fun with that fossil.
15. Sylvia Hope in Robertson et al. (2004, 766–67): "Listed here are major taxa (here used in the stem sense) of extant neornithine birds that are known in the Cretaceous from fossil records . . . or are assumed to have been present on the basis of ghost lineage requirements, or of well corroborated basal phylogeny and biogeography (phylogeography) that together indicate a Cretaceous evolutionary radiation."
16. Chiappe 2007, 238–41.
17. Chiappe 2007, 242.
18. Chiappe 2007, 245.
19. Holroyd, Wilson, and Hutchison 2009.
20. Much of the ecology of these extinct lizards is from my 1996 book and was provided by the work of my colleagues Richard Estes (1964) and Laurie Bryant (1989).
21. Matsumato and Evans (2009) provide the most recent overview of the group.
22. Erickson 1972.
23. Archibald 1996, 86–88.

24. Cvancara and Hoganson 1993.

25. Archibald 1996, table 5.1.

26. Brinkman, Newbrey, and Cook 2009.

27. Archibald 1996, table 5.1.

28. Carter and Wilson 2009, 331.

29. Archibald 1996, 172–78.

30. Johnson 1992, 117.

31. Wilf and Johnson 2004, 347.

32. Thompson, Arens, and Jahren 2009.

33. Dicotyledonous angiosperms are flowering plants that have two embryonic leaves, or cotyledons (e.g., most hardwood trees), compared to monocotyledonous angiosperms, which have a single cotyledon in the seed (e.g., lilies and grasses).

34. Nichols and Johnson 2008, 57.

35. Nichols and Johnson 2008, 131.

36. Iglesias et al. 2007.

37. Scholz and Hartman 2007.

Chapter 5: Causes of Extinction at the K/T Boundary

1. In one of my favorite *Far Side* cartoons (Larson and Martin 2003, 1: 601) a determined but dumb looking *T. rex*-like dinosaur and a beady-eyed mammal are both circling a large rock with writing implements in their hands (paws? claws?). The dinosaur has repeatedly printed, often with backward letters, "dinosaurs rule" while the mammal has repeatedly written in nice cursive "mammals rule." The punch line is that they have crossed out all of each other's messages. In the fertile mind of Gary Larson, mammals and nonavian dinosaurs may have competed.

2. Darwin 1859, 73–74. I would not of course advocate cats as a solution to other ecological imbalances.

3. Van Valen and Sloan 1966; Krause 1986.

4. Koenigswald and Schierning 1987.

5. Luo's (2007, 1017–18) claim that "developmental patterning and ecological diversification that were previously known only for Cenozoic mammals . . . are now shown to be widespread among Mesozoic mammals" is hyperbole. We now have some very interesting new finds showing limited ecological diversity among Mesozoic mammaliforms, but it is a pale comparison to the ecological diversity of mammals in the Cenozoic. In reference to the ecological diversity of Mesozoic mammals, one is reminded of Dorothy Parker's famous quote concerning a Katharine Hepburn performance, "She runs the gamut of emotions from A to B."

6. Archibald 1996, 116.

7. As Bill Clemens pointed out to me (pers. comm. 2009), he would add long-term climate change as a fourth testable cause. I agree that it almost certainly is a proximate cause, but I regard the three that I listed to be ultimate causes, any one of which could have caused long-term climate change.

8. In the realm of wacky hypotheses of nonavian dinosaur extinction, or in this case, nonextinction, is this new one provided by my departmental colleague Greg Harris, with tongue firmly in cheek. Supposedly nonavian dinosaurs were beamed-up by aliens just prior to the asteroid impact—shades of Douglas Adam's *Hitchhiker's Guide to the Galaxy* series, in which the dolphins depart Earth just before its destruction to make way for a hyperspace bypass, and *So Long, and Thanks for All the Fish* (Adams 1984).

9. There are a variety of opinions as to what struck Chicxulub 66 million years ago and what it should be called. The original Alvarez et al. paper (1980) called it an asteroid. It has also been re-ferred to as a *comet*, a *meteor*, a *meteorite*, a *chondritic projectile*, a *bolide*, or an *impactor*. As I am not an astrogeologist, I have no opinion. Where historically appropriate I term it an *asteroid*; otherwise I use neutral terminology, such as *projectile* or *impactor*.

10. Keller et al. (2004, 3753) argued, based on evidence from a drill core, that the Chicxulub crater "predated the K-T boundary by ~300,000 years and thus did not cause the end-Cretaceous mass extinction as commonly believed." Assuming that these authors are correct, it still begs the question: Where is the crater that produced the iridium, shocked quartz, and tektites known from the K/T boundary?

11. Chao et al. 1962, 419.

12. Hildebrand et al. 1991, 867.

13. Archibald and Fastovsky 2004, 684.

14. Arens and West 2008, 465.

15. Schulte et al. 2010, 1214.

16. Any number of Web sites used the hubristic term *dream team* in describing the group of 41 authors of the 2010 *Science* paper. This is one example: "A 'dream team' of researchers concludes that the massive Gulf of Mexico impact 65.5 million years ago—not volcanoes or multiple impacts—indeed caused the greatest extinction event of all. It's official: The extinction of the dinosaurs and a host of other species 65.5 million years ago was caused by a massive asteroid that crashed into the Gulf of Mexico, creating worldwide havoc, an international team of researchers said Thursday." http://poleshift.ning.com/profiles/blogs/confirmed-scientists-settle.

17. Archibald et al. 2010, 973.

18. Hildebrand et al. 1991, 867.

19. The Earth Impact Database www.unb.ca/passe/ImpactDatabase/index.html place the diameter of Chicxulub at 170 km.

20. Raup 1991, 167–79 and fig. 10–1.

21. Poag, Plescia, and Molzer 2002, 1100.

22. Raup 1991, 175.

23. The Earth Impact Database www.unb.ca/passc/ImpactDatabase/index.html.

24. Poag, Plescia, and Molzer 2002, 1095–96.

25. Deutsch and Koeberl 2006, 691.

26. Poag, Plescia, and Molzer 2002, 1081.

27. Poag, Plescia, and Molzer 2002, 1100.

28. See Paquay et al. 2008a (215–16 and references therein) for estimates of projectile size.

29. Morgan 2008, 1158a; Paquay et al. 2008b, 1158b.

30. Christian Koeberl, Department of Geological sciences seminar, "Meteorite impact cratering on earth: Geological and biological consequences," December 5, 2007.

31. Schultz and D'Hondt 1996, 963.

32. Schultz and D'Hondt 1996, 965–66.

33. Morgan et al. 2006, 264.

34. Morgan et al. 2006, 275–76.

35. Morgan et al. 2006, 278.

36. Alvarez et al. 1980, 1095, and 1106.

37. Pope 2002, 99.

38. Pope 2002, 101–2.

39. Alvarez et al. 1980, 1106.

40. www.igopogo.com/we_have_met.htm.

41. The pH value refers to the concentration of hydrogen ions (H+). Each unitary drop in the pH represents a tenfold increase in the concentration of hydrogen ions.

42. Cox 1993.

43. Lewis et al. 1982; Prinn and Fegley 1987; D'Hondt, Sigurdsson, et al. 1994, D'Hondt, Pilson, et al. 1994.

44. I quote from my 1996 book (139) concerning arguments for buffering by carbonate rocks,

> One strong advocate of K/T acid rain is Greg Retallack, a scientist studying ancient soils (e.g., 1993). When it has been pointed out to him that aquatic vertebrates, who would be most susceptible to ravages of acid rain, have among the best levels of K/T survival, he has countered that the surrounding soils or bedrock would have buffered the aquatic systems. He even suggested (1994, p. 1392), that "in such an acid rain crisis, limestone caves could have been important refugia for birds, mammals, amphibians, and small reptiles." The effects of acid rain can be buffered to some degree if the bedrock or soils are high in carbonates such as limestone. The problem with Greg's thesis is that there was no underlying carbonate bedrock in the latest Cretaceous of eastern Montana (or today, for that matter) that could have buffered the aquatic system. Further, because there was no carbonate bedrock underlying the aquatic systems, there could not have been any limestone caves in which creatures could have hidden. There are carbonate cements that commonly bind the uppermost Cretaceous sandstones together, but these are clearly derived from groundwaters percolating through the sediments long after the K/T boundary. Even when bedrock is dominated by carbonate rocks such as limestone, it does not insure against highly acidic water. Such is the case today for many lakes in Florida (Pollman and Canfield 1991).

45. Retallack, as recently as 2004, was still pushing for acid rain at the K/T boundary, but with a much-toned-down plea: "Similar limits come from a non-marine bioassay of pH suppression to less than 5.5 but no less than pH 4, allowing survival of amphibians and fish, but strong extinctions of non-marine mollusks in Montana" (35).

46. Nichols and Johnson 2008, 139–40.

47. Alvarez and Asaro 1990, 81.

48. Ectotherms, which include most vertebrates except birds, mammals, and a few fishes, obtain their heat from environmental sources, most often the sun. This is in contrast to birds, mammals, and the few fishes that obtain their heat through cellular metabolism.

49. Hutchison 1982.

50. Clemens and Nelms 1993.

51. Work on Arctic (and what amounts to Antarctic nonavian dinosaurs in Australia) vertebrates, especially nonavian dinosaurs, continues. An interesting *NOVA* program (Oct. 7, 2008) explored the topic. The program notes concluded: "Finally, the program touches on the ultimate implications of nonavian dinosaur survival. Did a catastrophic asteroid impact 65 million years ago wipe out the nonavian dinosaurs, as most people now believe, or did more gradual ecological changes play an equally decisive role in their demise? Like a good detective story, 'Arctic Nonavian dinosaurs' fingers new suspects in its search for answers to the extinction riddle, including massive volcanic eruptions, shifting continents, and a gradual climatic chill—the opposite of today's global warming." www.pbs.org/wgbh/nova/arcticdino/about.html.

52. See Gale (2006) regarding these tsunami deposits as well as the possibility of additional impacts in the Late Cretaceous.

53. Robertson et al. 2004a.

54. Archibald and MacLeod 2007, 2–3.

55. Varricchio, Martin, and Katsura 2007, 1361.

56. Martin 2009.

57. Belcher et al. 2003, 2004b; Robertson et al. 2004b. The argument for a global wildfire continues to be plagued with problems. Harvey et al. (2008, 255) write that

> minimal amounts of charred plant remains and abundant noncharred material occur in various [K/T] boundary locations across North America. This refutes the inference that wildfires occurred on a global scale, and requires an alternative explanation for the aciniform soot . . . [there are] significant concentrations of carbon cenospheres in [K/T] boundary sediments from New Zealand, Denmark, and Canada. Carbon cenospheres are thought to derive solely from incomplete combustion of pulverized coal or fuel-oil droplets, which suggests that the impact may have combusted organic-rich target crust. The Chicxulub impact crater is located adjacent to the Cantarell oil reservoir, one of the most productive oil fields on Earth. This indicates that abundance of organic carbon in the Chicxulub target crust was likely to have been above global mean values. But even if we discount Chicxulub's organic-rich locality, the global mean crustal abundance for fossil organic matter is more than adequate to account for observed concentrations of both carbon cenospheres and aciniform soot, therefore making the global wildfire hypothesis unnecessary.

58. Goldin and Melosh 2009.

59. In a "Research Focus" accompanying the Goldin and Melosh article, Claire Belcher (2009, 1148) concluded, "It is recognized that there was a major disruption to plant communities across the K-Pg boundary. . . . These new model-based results [referring to Goldin and Melosh, 2009], taken together with the abundant literature on paleontological indicators of fire occurrence, suggest that extensive wildfires were not the cause. This model-data agreement does not eliminate the role of relatively high temperatures (on the order of a couple of hundred degrees centigrade) in some of the extinctions seen at this time; but does suggest that the thermal pulse component of the K-Pg impact was not as significant as has been previously thought. Additional mechanisms might be required to fully explain the K-Pg extinctions."

60. Archibald 1996, 68.

61. We sadly lost Malcolm in 2008, at the age of 77.

62. James Fassett (e.g., 2009) is the most vocal proponent of Paleocene nonavian dinosaurs from sediments in the San Juan Basin of northwest New Mexico and southwestern Colorado.

63. Intertrappean sediments refer to sediments that lay between the layers of volcanic rocks known as traps. The word *trap* derives from Swedish, *trapp* or *trappa,* because of their stairlike appearance (Hoad 1996).

64. Keller et al. 2008.

65. Keller et al. 2008, 308.

66. Dewey McLean was an even earlier advocate of K/T boundary volcanism as a cause of mass extinction. His main argument for vertebrate extinctions never quite had traction. If I understood correctly, he thought greenhouse gases from the eruptions caused heat-induced environmental effects on embryo survival and population dynamics. It specifically involved environmental heat-induced reduction of blood flow to the uterine tract, damaging and killing embryos within their mothers (e.g., McLean 1995).

67. Archibald 1996, 143.

68. Thompson 2008.

69. Thompson (2008) notes, "For comparison, the 1991 Pinatubo eruption, which cooled Earth's climate for several years afterward, sent about 0.017 billion tons of sulfur dioxide into the atmosphere. The Chicxulub crater put anywhere between 50 billion to 500 billion tons of sulfur dioxide in the air. The entire Deccan traps spewed on the order of 10,000 billion tons in the atmosphere, Courtillot said."

70. www.nsf.gov/news/news_summ.jsp?cntn_id=111722.

71. Peters 2008, 626.

72. Dates are from Gradstein, Ogg, and Smith (2004) and are rounded upward to the nearest whole number.

73. Peters 2008, 627–28.

74. Raup and Sepkoski 1982, 1501, fig. 1.

75. Erwin 1993, 239.

76. Newell 1987, 306–7.

77. Newell (1967, 63–66) discusses the often-confusing way in which these concepts are treated.

78. The Cuvier passages come from the English translation by Ian Johnston based upon the third French edition (Paris and Amsterdam, 1825). The text is at www.scholars.nus.sg/landow/victorian/science/science_texts/cuvier/cuvier-e.htm.

79. Rudwick 1997, 262–63.

80. Nichols and Johnson 2008.

81. Nichols and Johnson 2008, 223–24.

82. Belt, Hicks, and Murphy 1997.

83. Nichols and Johnson (2008) do not show these tongues in their figure 6, but figure 5.3, taken from Scholz and Hartman, shows these units.

84. Bill Clemens pointed out a monograph in which R. W. Lemke (1960, 29) states that in North Dakota "the exposure near Verendrye indicates that the Hell Creek formation may be lacking in the eastern part of the area and that Cannonball strata may rest directly on Fox Hills sandstone. If so, both the Hell Creek formation and the Ludlow member of the Fort Union formation pinch out between Verendrye and the Des Lacs well." This is evident in his figure 4, in which he appears to be showing an onlap relationship in which the younger Cannonball overrides an erosional surface cut into the older beds (Hell Creek, Fox Hill, and Bearpaw/Pierre). This is generally associated with a marine transgression, as in this instance. Thus, in this case, while the younger marine Cannonball Formation may directly overlay the older, partially marine Fox Hills Sandstone, it probably does so unconformably, that is, the intervening terrestrial beds were either never deposited or were eroded away before the Cannonball Formation was deposited.

85. Gale 2006.

86. See Gallagher et al. 2005.

87. It was quite by accident that I became aware of the existence of completely freshwater stingrays. My wife and I traveled on the Madre de Dios and Manu rivers in the southwestern Amazon rainforests in the summers of 2007 and 2008. Our favorite guide on both trips, Marlene Huamán Berrocal, relayed a story of a young boatman who had been stung by a stingray while dangling his hand in the water. He had to be taken many miles on the river to reach medical treatment. After quickly withdrawing my hand into the boat, I queried her if she meant the flat, winged relatives of sharks—and indeed she did. Never mind the large caimans, piranhas, or electric eels, the stingrays (three species Paratrygon aiereba, Potamotrygon castexi, and Potamotrygon motoro) can reach a meter across. The waterfalls on the Madre de Dios River in Bolivia, which have kept the pink river dolphin from reaching the upper part of the same drainage in southwestern Peru, may have kept the stingrays isolated upstream.

88. Hoganson and Murphy 2002, 267.

89. Sankey (2009, 126–27) reported a screen-washed "channel deposit 29.9 m below the K/T and ~65.9 Ma and . . . a crevasse splay 8.4 m below the K/T and ~65.6 Ma." These yielded far more skate and ray teeth than by surface collecting alone. She argues that this "indicates the presence of a seaway in the area just prior to the K/T." She does not indicate her age definition for the K/T boundary, but using her stratigraphic positions and age estimates, there was a minimum of 100,000 years between her argument for a seaway and the end of the Cretaceous, after which a seaway is not known locally.

90. Hoganson and Murphy 2002, 267. The one species that they note as possibly extending back into the Late Cretaceous is Otodus obliquus, which is regarded as post-K/T boundary only by most other workers.

91. Wilf, Johnson, and Huber 2003, 599.

92. Wilson, http://protist.biology.washington.edu/GPWilson/Mammalian_change.htm.

93. Wilson 2009, 150.

94. Habib and Saeedi 2007, 87.

95. Thompson, Arens, and Jahren 2009, 151.

96. This scenario is based on the discussions in this chapter and is modified after Archibald 1996 and Archibald and Fastovsky 2004.

97. Another group of placental mammals, Xenarthra (sloths, armadillos, anteaters), or their immediate ancestors reached South America by at least the K/T boundary if not before. While they reached North America much later, they are otherwise a uniquely South American group.

Chapter 6: After the Impact: Modern Mammals, When and Whence

1. The science of building cladograms (clade means "branch") or phylogenies is known as phylogenetic systematics or cladistics. It attempts to recover the evolutionary history of life. The question is of course how to do this. Cladistics argues that one should use only derived characters or traits that are shared through common ancestry to argue for evolutionary relationships. While ancestral or primitive characters or traits are biologically interesting, as are traits that have evolved because of similar ecological constraints, neither can elucidate evolutionary relationships. A classic case is the origin of flight and the modifications of structures that it requires. Bats, birds, and pterosaurs all have wings for powered flight, but because of the underlying structure of their wings we can be quite sure that they evolved flight separately.

Although greatly modified, all three flighted groups have the telltale signs of the ancestral tetrapod limb within their wing— a humerus, a radius, an ulna, wrist bones, and up to five digits. Here the similarities end. While the bat's wings are composed of parts of all five fingers, that of the bird is composed of the entire forelimb and that of the pterosaur is composed of an elongated fourth finger. Thus, while the underlying possession of tetrapod limb bones shows that these three groups shared a common ancestor with other tetrapods (specifically other amniotes), the varying structures composing the wing show that ecological constraints were imposed by three separate origins of flight in these three vertebrate groups (convergent evolution in the three groups). Dealing with such obvious differences seems easy to us today, but it was not the case, for example, when pterosaurs were first described. Although some scientists, such as the turn-of-the-nineteenth-century paleontologist and anatomist Georges Cuvier, realized that pterosaurs were reptiles, others lacked such an anatomical appreciation, variously placing pterosaurs with bats or birds. Thanks to powerful desktop computers, such analyses have become more sophisticated as well as easier to perform.

2. The source for the various numbers of species of mammals reported in this section are from Wilson and Reeder 2005, table 1.

3. Not counting marine species, there are about 140 species of native bats and native rodents in Australia according to Strahan 2000. This estimate may be somewhat low.

4. Pascual et al. 2000.

5. Krause (2001) described a partial lower molar from the Late Cretaceous of Madagascar as belonging to a marsupial. Averianov et al. (2003) suggested it may instead belong to the eutherian group Zhelestidae. Whether the tooth is metatherian or eutherian, it almost certainly belongs to a therian of some sort. Also Prasad and his colleagues have described what are almost certainly eutherians from the Late Cretaceous of India (Godinot and Prasad 1994; Prasad et al. 1994; Prasad and Sahni 1988; Prasad et al. 2007). While India was certainly of Gondwanan origin, there is a question as to when it may have started exchanging faunal and floral elements with Laurasia.

6. The dates of 166 million years ago for the origin of mammals and the split between Prototheria and Theria, and 148 million years ago for the split between Metatheria and Eutheria, are based on the molecular phylogentic study of Bininda-Emonds et al. 2007. Molecular evidence suggests as recently as 21.2 million years ago for the split between the platypus and echidna lineages (Warren et al. 2008), but fossil evidence suggests a platypus between 121 and 112.5 million years and thus a platypus and echidna split by then (Rowe et al. 2008). The first group of authors suggests that this could "perhaps be owing to relatively recent reductions of mutational rates in the monotreme lineage."

7. I note that Kielan-Jaworowska, Cifelli, and Luo (2004) use *Mammalia* for I what I have called *Mammaliaformes.*

8. Bininda-Emonds et al. 2007.

9. Ji et al. (2002) named their new eutherian taxon *Eomaia,* literally meaning "dawn mother." Luo et al. (2003) called their new metatherian taxon *Sinodelphys,* a name that clearly indicates the authors' belief in the affinities of this creature. *Sinodelphys* literally means "Chinese womb," but *delphys* is a common term used in marsupial names. For example, the name *Didelphis,* the common North American opossum, refers to the two-part reproductive system characteristic of marsupials.

10. Vullo et al (2009) reported on teeth from the early Late Cretaceous of southwestern France, which were referred to a new primitive marsupial-like form.

11. Pers. comm., Alexander Averianov 2009.

12. Muizon and Argot 2003; Muizon and Cifelli 2000.

13. Godthelp et al. 1992; Beck et al. 2008.

14. Most of the preceding discussion relies on fossil and molecular dates and phylogenies found in Meredith et al. 2008.

15. Murphy et al. 2001, 2003.

16. The unfortunate practice of naming every new molecularly proposed taxon has become a problem in communication. Asher, Bennett, Lehmann (2009, 853) noted, "A variety of names for subclades within the new placental mammal tree have been proposed, not all of which follow conventions regarding priority and stability." Earlier, I (Archibald 2003, 356–57) had written:

In some recent molecular studies there has been an unfortunate tendency to name even slightly modified ordinal or superordinal clades. The International Code of Zoological Nomenclature (1999, fourth edition) does not specifically address taxa above the familial level. For the levels it does cover, however, one does not in most cases apply a completely new name each time there are modifications to the contents of an existing taxon, thus there is rationale for retaining old names while in other cases new names seem justified. For example, if in a study two taxa are newly found to be each other's sister taxon, then a new taxon is appropriate such as Cetartiodactyla, if Cetacea and Artiodactyla are sister taxa. Most molecular studies now place Cetacea within Artiodactyla probably as sister taxon to Hippopotamidae, thus a new name for these sister taxa is warranted, such as the whimsical Whippomorpha (Gatesy et al. 1996; Waddell et al. 1999b) but not a new name for Artiodactyla. Similarly, as it became ever clearer that Aves was within Dinosauria, the latter name was retained rather than naming a new taxon Avedinosauria. Similarly, if a taxon loses more inclusive taxa, there is no rationale for applying a new name. Hence, because Lipotyphla has lost Chrysochloridae and Tenrecidae to a new order Tenrecoidea (McDowell 1958; Afrosoricida of Stanhope et al., 1998), it does not follow that the former taxon now should be Eulipotyphla (Waddell et al. 1999a,b). The same rationale pertains to rejecting Euarchonta in favor of Archonta, Fereuungulata in favor of Ferungulata (Waddell et al. 1999a,b). It is reasonable to apply new names such as Laurasiatheria (Waddell et al., 1999b), Euarchontoglires (Murphy et al. 2003), Afrotheria (Stanhope et al. 1998), and Boreoeutheria (Murphy et al. 2003) (whether one finds them euphonious or not) that pertain to essentially new clades.

17. As can be seen in figure 6.1, Xenarthra appears to be a superordinal group composed only of Xenarthra. This is more of a historical accident than of biological consequence. The classificatory rank or category "order," like all others in the taxonomic hierarchy, has no biological meaning, except perhaps for "species," which is the level within which evolution operates. These names are simply placeholders for the biologically real entities, the taxa (taxon). For example, *order* helps place but does not define *Carnivora.* In the past, the ordinal rank was the most inclusive we could attain in assessing relationships, and interestingly, as I discuss, they all arose within about the first 15 million years of the Cenozoic, suggesting some ecological connection. Increasingly, two orders of Xenarthra are recognized: Cingulata, for armadillos, and Pilosa, for anteaters and sloths. I endorse these recommendations and also note that Cingulata and Pilosa first appear in the 15-million-year window in the early Cenozoic, albeit in South America.

18. Simpson 1945.

19. Simpson 1945 (Ferungulata) and Gregory 1910 (Archonta).

20. Foote et al. 1999.

21. Kirchner and Weil (2000) make a case, using the entire Phanerozoic marine fossil record, that there is a lag time of about 10 million years between mass extinctions and origination rates related to recoveries. For mammals it is probably fair to say that a complete recovery after the K/T boundary did not occur until near the beginning of the Eocene (about 10 million years later). Nevertheless, the fossil record, at least for North America, shows a huge spike in mammalian originations within a million years

and probably considerably less. While many of these originations were dominated by condylarths, there was a diversity of other sizes and dietary preferences, especially by at least the Late Paleocene.

22. Not all other studies have found the same trends or patterns; for example, Alroy (2009) did not find the same trends or at least at the same magnitude for North America Cenozoic mammals.

23. Kennett et al. 1974.

24. Wilson 2005.

25. See chap. 4, n. 1.

26. One of the questions that is only now being addressed is from whom and whence do the new species in the Paleocene in the Western Interior come? The "from whom" part of the question implies some sort of speciation events within the Western Interior, while the "whence" part asks where, if not from the Western Interior, did the other new species originate? These are difficult questions requiring considerable phylogenetic and biogeographic data, which we for the time being do not have. On a smaller scale, Bill Clemens and former students Anne Weil and Greg Wilson have looked at this interval for the Early Paleocene in northeastern Montana (Clemens 2002; Weil and Clemens 1998). Based on table 1 in Clemens (2002), at the earliest Paleocene sites five species are "residents," meaning that they are found earlier in the Late Cretaceous or are derived from Late Cretaceous species in the region. In the same sites, 12 are "aliens," meaning that they originated elsewhere. In the next younger but still Early Paleocene sites there are now 17 residents, 15 aliens, and four unknowns. These numbers include all major groups of mammals—multituberculates, metatherians (although greatly reduced), and eutherians. With a relative increase in residents, it seems that there is a shift away from more strictly immigration to in situ evolution.

27. According to John Alroy (pers. comm. 2009) the "other" in his figure (here fig. 6.5) includes *Arctostylops*, Apatemyidae, Pantolestidae, Leptictidae, Taeniodonta, Tillodontia, Pantodonta, Dinocerata, Palaeanodonta, Dermoptera, Pholidota, Proboscidea, Toxodontidae, and Xenarthra as well as eutherian taxa in the Late Cretaceous. The figure excluded volant and marine groups (Chiroptera, Pinnipedia, Desmostylia, Sirenia, and Cetacea).

28. One of my anonymous reviewers complained that I should not use the nineteenth-century phrase *nonadaptive radiation*. While I share the reviewer's dislike of the term, as I noted in the text, there is a body of newer literature on the topic. The term is here to stay, at least for now. In 1990, Gittenberger asked the question, "What about non-adaptive radiation?" He noted that the expression *adaptive radiation* suggests that the term *radiation* alone seems to imply a lack of adaptation. This can lead to issues when trying to examine the radiations of related taxa. He examined land snails in Greece and surrounding countries to explore this process and its resulting pattern. In his conclusion he stated (271), "If radiation occurs, this will be more or less clearly adaptive, non-adaptive or non-adaptive changing into adaptive while the process proceeds." The last of these three is exactly the pattern I am exploring in this chapter, but on a longer and larger scale. There are other explorations of nonadaptive versus adaptive radiations. Sanderson 1998 presented a succinct review of concepts surrounding adaptive and nonadaptive radiations. Kozak, Weisrock, and Larson (2006) dealt with the rapid appearance of lineages of woodland salamanders (Plethodontidae: *Plethodon*) showing relatively little ecological diversification (nonadaptive). Sidlauskas (2008) used a group of South American bony fish (Characiformes: Anostomoidea) to examine why in one clade there is great morphologic diversification and the other there is not. Finally, Rundell and Price (2009) reviewed how nonecological and ecological speciation may be related or even drive nonadaptive and adaptive radiations, respectively.

29. Exceptions, which certainly included some weird forms, were the 20 or so genera of pantodonts, dinoceratans, taeniodonts, and tillodonts known from the Paleocene of North America and elsewhere. Sizes ranged from smaller dog-sized to long-legged rhinoceros-sized, and ecologically from specialized diggers to very large herbivores with jaws armed with large canines and many horned heads. They declined in the Eocene and did not give rise to any modern groups of mammals.

30. Fox (1989) has reported the only instances of condylarths that well may be from the Late Cretaceous. As discussed by Lofgren (1995), all other early reports, notably from eastern Montana in general and the Bug Creek area in particular, are definitely or most likely earliest Paleocene rather than latest Cretaceous in age. The newer, rarer finds from both northeastern and southeastern Montana discussed earlier may change this assessment.

31. Archibald 1998.

32. Muizon and Argot 2003; Muizon and Cifelli 2000.

33. Godthelp et al. 1992; Beck et al. 2008.

34. Patton et al. 2000.

35. Wible et al. 2007.

36. Archibald and Deutschman 2001.

37. The molecular dates come from Bininda-Emonds et al. (2007) and are here rounded to the nearest million years.

38. The only published report of which I am aware claiming a Cretaceous rodent was a newspaper article in the *Hindu*, dated Friday, November 2005, and titled "World's oldest fossil of rodent discovered in Jaisalmer." It was claimed to have come from a span ranging from 93.2 million to 142 million years ago. Needless to say, nothing else has been heard of this great find in the scientific literature. The fossil of the rodent's tooth forms part of a microvertebrate fossil assemblage unearthed in the Jaisalmer district of Rajasthan.

39. Hasegawa et al. 2003; Springer et al. 2003.

40. Benton and Donoghue 2007; Benton et al. 2009.

41. Hug and Roger 2007.

42. Bromham et al. 1996.

43. Goodman et al. 2009.

44. Omland 1997.

45. Bromham 2009, 1: "Despite hopes that the processes of molecular evolution would be simple, clock-like and essentially universal, variation in the rate of molecular evolution is manifest at all levels of biological organization. Furthermore, it has become clear that rate variation has a systematic component: rate of molecular evolution can vary consistently with species body size, population dynamics, lifestyle and location. This suggests that the rate of molecular evolution should be considered part of life-history variation between species, which must be taken into account when interpreting. DNA sequence differences between lineages. Uncovering the causes and correlates of rate variation may allow the development of new biologically motivated models of molecular evolution that may improve bioinformatic and phylogenetic analyses."

46. Bromham and Woolfit (2004, 758) found "no evidence of a consistent increase in rates in island taxa compared to their mainland relatives, and therefore [they] find no support for the hypothesis that the molecular clock runs fast in explosive radiations." In a

more recent article Fontanillas et al. (including Bromham) (2007, 7) indicated: "Our results also suggest that molecular dates might be overestimated when larger animals are over-represented in the sampled taxa. Such a bias could apply even if the body size of Metazoa has not systematically increased. It is also important to note that our results apply only to mitochondrial genes, yet Precambrian molecular dates have predominantly been generated from nuclear gene data."

47. Murphy et al. 2005, 614. These authors used Springer et al. 2003 for the times of divergence for the various clades. The cladogram in Springer et al. 2003 has the same topology as that in figure 6.1.

48. The same sources cited by Murphy et al., 2005, caption to fig. 3 were used in the new calculations in Fig. 6.9 except that for the timing of the origin of placental orders, the date of 66 million years ago, the K/T boundary, was used.

49. Asher et al. 2009, 385.

50. Murphy et al. 2003.

51. Hunter and Janis 2006a,b.

52. Nishihara et al. 2009.

53. Krause 2001.

54. Godinot and Prasad 1994; Prasad et al. 1994; Prasad and Sahni 1988; Prasad et al. 2007.

55. Godthelp et al. (1992) made a single tooth from the Eocene Tingamarra local fauna in Australia the type and only known tooth of the genus *Tingamarra*. They referred it to as the eutherian paraphyletic group "Condylartha." It is the only non-bat eutherian to be claimed for such early Tertiary beds in Australia. The rat/mouse family Muridae includes the only terrestrial placentals known with certainty to have reached Australia before humans arrived. They are still there.

Epilogue: Lessons from the Past

1. Most of the story comes from Mac Guckin de Slane's 1845 translation of Ibn Khallikān's *Biographical Dictionary*, 71–73, but the final calculations are taken from the much more comprehensible Web site, Kidipede, www.historyforkids.org/learn/islam/literature/chesswheat.htm, which tells a similar story, "The Story of Wheat on the Chessboard." A children's version of this tale, by David Birch, identifies the king as coming from Deccan, India, which is rather apropos for a book on K/T boundary extinctions.

2. U.S. Census Bureau, International Programs, www.census.gov/ipc/www/popclockworld.html.

3. Pennisi 2008, 179.

4. Schipper et al. 2008, 228–29.

REFERENCES

Adams, D. 1984. *So Long, and Thanks for All the Fish*. London: Pan Books.

Alcalá, L., and R. Royo-Torres. (coord.) 2009. Mesozoic Terrestrial Ecosystems in Eastern Spain. *¡Fundamental!* 14:1–153.

Alroy, J. 2009. Speciation and extinction in the fossil record of North American mammals. In R. K. Butlin, J. R. Bridle, and D. Schulter, eds., *Speciation and Patterns of Diversity*, 301–23. Cambridge: Cambridge University Press.

Alvarez, L. W., W. Alvarez, F. Asaro, and H. V. Michel. 1980. Extraterrestrial cause for the Cretaceous-Tertiary extinction. *Science* 208:1095–1108.

Alvarez, W., and F. Asaro. 1990. An extraterrestrial impact. *Scientific American* 263:78–84.

Archibald, J. D. 1982. A study of Mammalia and geology across the Cretaceous-Tertiary boundary in Garfield County, Montana. *University of California Publications in the Geological Sciences* 122:1–286.

———. 1996. *Dinosaur Extinction and the End of an Era: What the Fossils Say*. New York: Columbia University Press.

———. 1998. Archaic ungulates ("Condylartha"). In C. Janis, K. Scott, and L. Jacobs, eds. *Evolution of Tertiary Mammals of North America*. Vol. 1, *Terrestrial Carnivores, Ungulates, and Ungulatelike Mammals*, 292–331. Cambridge: Cambridge University Press.

———. 2003. Timing and biogeography of the eutherian radiation: Fossils and molecules compared. *Molecular Phylogenetics and Evolution* 28:350–59.

———. 2006. A cacophony of causes. Review of *Extinction: How Life Nearly Ended 250 Million Years Ago*, by Douglas H. Erwin. *Trends in Ecology and Evolution* 21:428.

Archibald, J. D., and A. O. Averianov. 2001. *Paranyctoides* and allies from the Late Cretaceous of North America and Asia. *Acta Palaeontologica Polonica* 46:533–51.

———. 2005. Mammalian faunal succession in the Cretaceous of the Kyzylkum Desert. *Journal of Mammal Evolution* 12:9–22.

———. 2006. Late Cretaceous asioryctitherian eutherian mammals from Uzbekistan and phylogentic analysis of Asioryctitheria. *Acta Palaeontologca Polonica* 51:351–76.

Archibald, J. D., A. O. Averianov, and E. G. Ekdale. 2001. Late Cretaceous relatives of rabbits, rodents, and other extant eutherian mammals. *Nature* 414:62–65.

Archibald, J. D., and D. Deutschman. 2001. Quantitative analysis of the timing of origin of extant placental orders. *Journal of Mammalian Evolution* 8:107–24.

Archibald, J. D., and D. E. Fastovsky. 2004. Dinosaur extinction. In Weishampel, Dodson, and Osmólska, *The Dinosauria*, 672–84.

Archibald, J. D., and N. MacLeod. 2007. S.v. "Dinosaurs, extinction theories for." *Encyclopedia of Biodiversity*, 1–9. Elsevier, http://doi:10.1016/B0-12-226865-2/00071-7.

Archibald, J. D., W. A. Clemens, K. Padian, and 26 others. 2010. Cretaceous extinctions: Multiple causes. *Science* 328:973.

Arens, N. C., and I. D. West. 2008. Press-pulse: A general theory of mass extinction? *Paleobiology* 34:456–71.

Asher, R. J. 2005. Insectivoran-grade placentals. In Rose and Archibald, *The Rise of Placental Mammals*, 50–70.

Asher, R. J., N. Bennett, and T. Lehmann. 2009. The new framework for understanding placental mammal evolution. *Bioessays* 31:853–64.

Asher, R. J., J. Meng, J. R. Wible, M. C. McKenna, G. W. Rougier, D. Dashzeveg, and M. J. Novacek. 2005. Stem Lagomorpha and the antiquity of Glires. *Science* 307:1091–94.

Averianov, A. O., and J. D. Archibald. 2005. Mammals from the mid-Cretaceous Khodzhakul Formation, Kyzylkum Desert, Uzbekistan. *Cretaceous Research* 26:593–608.

Averianov, A. O., J. D. Archibald, and T. Martin. 2003. Placental nature of the alleged marsupial from the Cretaceous of Madagascar. *Acta Palaeontologica Polonica* 48:149–51.

Bailey, J. V., A. S. Cohen, and D. A. Kring. 2005. Lacustrine fossil preservation in acidic environments: Implications of experimental and field studies for the Cretaceous-Paleogene Boundary Acid Rain Trauma. *PALAIOS* 20:376–89.

Barrett, P. M., A. J. McGowan, and V. Page. 2009. Dinosaur diversity and the rock record. *Proceedings of the Royal Society B* 276:2667–74.

Beck, R. M. D., H. Godthelp, V. Weisbecker, M. Archer, and S. J. Hand. 2008. Australia's oldest marsupial fossils and their biogeographical implications. *PLoS ONE* 3:e1858, http://doi:10.1371/journal.pone.0001858.

Belcher, C. M. 2009. Reigniting the Cretaceous-Paleogene firestorm debate. *Geology* 37:1147–48.

Belcher, C. M., M. E. Collinson, A. R. Sweet, A. R. Hildebrand, and A. C. Scott. 2003a. Fireball passes and nothing burns—

the role of thermal radiation in the Cretaceous-Tertiary event: Evidence from the charcoal record of North America. *Geology* 31:1061–64.

———. 2003b. Fireball passes and nothing burns—the role of thermal radiation in the Cretaceous-Tertiary event: Evidence from the charcoal record of North America: Reply. *Geology* 32:e50, http://doi:10.1130/G19989.1.

Belt, E. S., J. F. Hicks, and D. A. Murphy. 1997. A pre-Lancian regional unconformity and its relationships to Hell Creek paleogeography in southeastern Montana. *Contributions to Geology, University of Wyoming* 31:1–26.

Benton, M. J., and C. J. Donoghue. 2007. Palentological evidence to date the tree of life. *Molecular Biology and Evolution* 24:26–53.

Benton, M. J., P. C. J. Donoghue, and R. J. Asher. 2009. Calibrating and constraining molecular clocks. In S. B. Hedges and S. Kumar, eds., *The Time Tree of Life*, 35–86. Oxford: Oxford University Press.

Bininda-Emonds, O. R. P., M. Cardillo, K. E. Jones, R. D. E. MacPhee, R. M. D. Beck, R. Grenyer, S. A. Price, R. A. Vos, J. L. Gittleman, and A. Purvis. 2007. The delayed rise of present-day mammals. *Nature* 446:507–12.

Birch, D. 1993. *The King's Chessboard*. New York: Puffin Books.

Brinkman, D. B., M. Newbrey, and T. Cook. 2009. Fish of the Hell Creek Formation. *North American Paleontological Convention (NAPC 2009): Abstracts*, p. 142. www.napc2009.org/technical-program-and-abstracts.

Brochu, C. A. 1997. A review of *"Leidyosuchus"* (Crocodyliformes, Eusuchia) from the Cretaceous through Eocene of North America. *Journal of Vertebrate Paleontology* 17:679–97.

———. 2003. Phylogenetic approaches toward crocodylian history. *Annual Review of Earth and Planetary Sciences* 31:357–97.

Bromham, L. 2009. Why do species vary in their rate of molecular evolution? *Biology Letters* 5:401–4.

Bromham, L., A. Rambaut, and P. H. Harvey. 1996. Determinants of rate in mammalian DNA sequence evolution. *Journal of Molecular Evolution* 43:610–21.

Bromham, L., and M. Woolfit. 2004. Explosive radiations and the reliability of molecular clocks: Island endemic radiations as a test case. *Systematic Biology* 53:758–66.

Bryant, L. J. 1989. Non-dinosaurian lower vertebrates across the Cretaceous-Tertiary boundary in northeastern Montana. *University of California Publications in Geological Sciences* 134:1–107.

Buffetaut, E. 2003. *La fin des dinosaures (The End of the Dinosaurs)*. Paris: Fayard.

Carter, G. E., and G. P. Wilson. 2009. Amphibian paleocommunity dynamics of the Hell Creek Formation in northeastern Montana and the Cretaceous-Tertiary extinction event. *North American Paleontological Convention (NAPC 2009): Abstracts*, p. 331. www.napc2009.org/technical-program-and-abstracts.

Chao, E. C. T., J. J. Fahey, J. Littler, and D. J. Milton. 1962. Stishovite, SiO2, a very high pressure new mineral from Meteor Crater, Arizona. *Journal of Geophysical Research* 67:419–21.

Chester, S. G. B., E. J. Sargis, F. S. Szalay, J. D. Archibald, and A. O. Averianov. 2008. Therian femora from the Late Cretaceous of Uzbekistan. *Journal of Vertebrate Paleontology* 28 (supplement to no. 3):53A.

———. 2010. Mammalian distal humeri from the Late Cretaceous of Uzbekistan. 2010. *Acta Palaeontologica Polonica* 55:199–211.

Chiappe, L. M. 2007. *Glorified Dinosaurs: The Origin and Early Radiation of Birds*. Hoboken, NJ: Wiley.

Chiappe, L. M., and L. M. Witmer. 2002. *Mesozoic Birds: Above the Heads of Dinosaurs*. Berkeley: University of California Press.

Cifelli, R. L. 1990. Cretaceous mammals of southern Utah. IV. Eutherian mammals from the Wahweap (Aqulian) and Kaiparowits (Judithian) formations. *Journal of Vertebrate Paleontology* 10:346–60.

———. 1994. Therian mammals of the Terlingua local fauna (Judithian), Aguja Formation, Big Bend of the Rio Grande, Texas. *Contributions to Geology, University of Wyoming* 30:117–36.

———. 1999. Tribosphenic mammal from the North American Early Cretaceous. *Nature* 401:363–66.

Cifelli, R. L., J. J. Eberle, D. L. Lofgren, J. A. Lillegraven, and W. A. Clemens. 2004. Mammalian biochronology of the latest Cretaceous. In M. O. Woodburne, ed., *Late Cretaceous and Cenozoic Mammals of North America*, 21–42. New York: Columbia University Press.

Clemens, W. A. 1964. Fossil mammals of the type Lance Formation, Wyoming: Part I. Introduction and Multituberculata. *University of California Publications in Geological Sciences* 48:1–105.

———. 1966. Fossil mammals of the type Lance Formation, Wyoming: Part II. Marsupialia. *University of California Publications in Geological Sciences* 62:1–122.

———. 1973. Fossil mammals of the type Lance Formation, Wyoming: Part III. Eutheria and Summary. *University of California Publications in Geological Sciences* 94:1–102.

———. 1980. *Gallolestes pachymandibularis* (Theria, *incertae sedis*; Mammalia) from the Late Cretaceous deposits in Baja California del Norte, Mexico. *PaleoBios* 33:1–10.

———. 2002. Evolution of the mammalian fauns across the Cretaceous-Tertiary boundary in northeastern Montana and other areas of the Western Interior. In Hartman, Johnson, and Nichols, *The Hell Creek Formation and the Cretaceous-Tertiary Boundary in the Northern Great Plains*, 217–45.

Clemens, W. A., and L. G. Nelms. 1993. Paleoecological implications of Alaskan terrestrial vertebrate fauna in latest Cretaceous time at high paleolatitudes. *Geology* 21:503–6.

Cope, E. D. 1881. A new type of Perissodactyla. *American Naturalist* 15:1017–2018.

Courtillot, V. E. 1990. A volcanic eruption. *Scientific American* 263:85–92.

Cox, G. W. 1993. *Conservation Ecology Biosphere and Biosurvival*. Dubuque, IA: Wm. C. Brown.

Crompton, A. W., D. E. Lieberman, T. Owerkowicz, R. V. Baudinette, and J. Skinner. 2008. Motor control of masticatory movements in the Southern hairy-nosed wombat (*Lasiorhinus latifrons*). In C. Vinyard, M. J. Ravosa, and C. Wall, eds., *Primate Craniofacial Function and Biology*, 83–111. New York: Springer Verlag.

Currie, P. J. 2003. Allometric growth in tyrannosaurids (Dinosauria: Theropoda) from the Upper Cretaceous of North America and Asia. *Canadian Journal of Earth Sciences* 40:651–65.

———. 2005. Theropods, including birds. In Currie and Koppelhus, *Dinosaur Provincial Park*, 367–97.

Currie, P. J., and E. B. Koppelhus. 2005. *Dinosaur Provincial Park: A Spectacular Ancient Ecosystem Revealed*. Bloomington: Indiana University Press.

Cuvier, G. 1825. Discours sur les révolutions de la surface du globe, et sur les changemens qu'elles ont produits dans le règne animal [Discourse on the Revolutionary Upheavals on the Surface of the Globe and on the Changes which They Have Produced]. 3rd ed. Paris: Dufour et d'Ocagne.

Cvancara, A. M., and J. W. Hoganson. 1993. Vertebrates of the Cannonball Formation (Paleocene) in North and South Dakota. *Journal of Vertebrate Paleontology* 13:1–23.

Darwin, C. R. 1859. *On the Origin of Species by Means of Natural Selection; or, The Preservation of Favoured Races in the Struggle for Life.* London: John Murray.

Desmond, A. 1982. *Archetypes and Ancestors Paleontology in Victorian London, 1850–1875.* Chicago: University of Chicago Press.

———. 1984. Interpreting the origin of mammals: New approaches to the history of palaeontology. *Zoological Journal of the Linnean Society* 82:7–16.

Deutsch, A., and C. Koeberl. 2006. Establishing the link between the Chesapeake Bay impact structure and the North American tektite strewn field: The Sr-Nd isotopic evidence. *Meteoritics and Planetary Science* 41:689–703.

D'Hondt, S., H. Sigurdsson, A. Hanson, S. Carey, and M. Pilson. 1994. Sulfate volatilization, surface-water acidification, and extinction at the KT boundary. New developments regarding the KT event and other catastrophes in earth history. *Lunar and Planetary Institute Contribution,* no. 825:29–30.

D'Hondt, S., M. E. Q. Pilson, H. Sigurdsson, A. K. Hanson Jr., and S. Carey. 1994. Surface water acidification and extinction at the Cretaceous-Tertiary boundary. *Geology* 22:983–86.

Dingus, L., and T. Rowe. 1997. *The Mistaken Extinction.* New York: W. H. Freeman and Co.

Eberth, D. 2005. The geology. In Currie and Koppelhus, *Dinosaur Provincial Park,* 54–82.

Ekdale, E. 2008. Variation among endocasts of the bony labyrinth of zhelestids (Mammalia: Eutheria). *Journal of Vertebrate Paleontology* 28 (supplement to no. 3):74A.

Erickson, B. R. 1972. The lepidosaurian reptile *Champsosaurus* in North America. *Monograph of the Science Museum of Minnesota* 1:1–91.

Erwin, D. H. 1993. *The Great Paleozoic Crisis: Life and Death in the Permian.* New York: Columbia University Press.

———. 2006. *Extinction: How Life on Earth Nearly Ended 250 Million Years Ago.* Princeton: Princeton University Press.

Estes, R. 1964. Fossil vertebrates from the Late Cretaceous Lance Formation, eastern Wyoming. *University of California Publications in Geological Sciences* 49:1–187.

Fassett, J. E. 2009. New geochronologic and stratigraphic evidence confirms the Paleocene age of the dinosaur-bearing Ojo Alamo Sandstone and Animas Formation in the San Juan Basin, New Mexico and Colorado. *Palaeontologia Electronica* 12(1):3A.

Fastovsky, D. E., Y. Huang, J. Hsu, J. Martin-McNaughton, P. M. Sheehan, and D. B. Weishampel. 2004. The shape of Mesozoic dinosaur richness. *Geology* 32:877–80.

———. 2005. The shape of Mesozoic dinosaur richness: Reply. *Geology* Online Forum, e75, http://doi:10.1130/G20695.1.

Feduccia, A. 1995. Explosive evolution in Tertiary birds and mammals. *Science* 267:637–38.

———. 1996. *The Origin and Evolution of Birds.* New Haven: Yale University Press.

Flynn, J. J., and G. D. Wesley-Hunt. 2005. Carnivora. In Rose and Archibald, *The Rise of Placental Mammals,* 175–98.

Flynn, L. J. 1986. Late Cretaceous mammal horizons from the San Juan Basin, New Mexico. *American Museum Novitates* 2845:1–30.

Fontanillas, E., J. J. Welch, J. A. Thomas, and L. Bromham. 2007. The influence of body size and net diversification rate on molecular evolution during the radiation of animal phyla. *BioMed Central Evolutionary Biology* 7:95.

Foote, M., J. P. Hunter, C. M. Janis, and J. J. Sepkoski Jr. 1999. Evolutionary and preservational constraints on origins of biologic groups: Divergence times of eutherian mammals. *Science* 283:1310–14.

Fox, R. C. 1970. Eutherian Mammal from the early Campanian (Late Cretaceous) of Alberta, Canada. *Nature* 227:630–31.

———. 1971. Early Campanian multituberculates (Mammalia: Allotheria) from the Upper Milk River Formation, Alberta. *Canadian Journal of Earth Sciences* 8:916–38.

———. 1976. Additions to the mammalian local fauna from the upper Milk River Formation (Upper Cretaceous), Alberta. *Canadian Journal of Earth Sciences* 13:1105–18.

———. 1979a. Mammals from the Upper Cretaceous Oldman Formation, Alberta. I. *Alphadon* Simpson. *Canadian Journal of Earth Sciences* 16:91–102.

———. 1979b. Mammals from the Upper Cretaceous Oldman Formation, Alberta. II. Pediomys Marsh (Marsupialia). *Canadian Journal of Earth Sciences* 16:103–13.

———. 1979c. Mammals from the Upper Cretaceous Oldman Formation, Alberta. III. Eutheria. *Canadian Journal of Earth Sciences* 16:114–25.

———. 1984. *Paranyctoides maleficus* (new species), an early eutherian mammal from the Cretaceous of Alberta. *Carnegie Museum of Natural History, Special Publication* 9:9–20.

———. 1989. The Wounded Knee Local Fauna and mammalian evolution near the Cretaceous-Tertiary boundary, Saskatchewan, Canada. *Palaeontographica, Abteilung A* 208:11–59.

Fox R. C., and B. G. Naylor. 2003, A Late Cretaceous taeniodont (Eutheria, Mammalia) from Alberta, Canada. *Neues Jahrbuch für Geologie und Paläontologie, Abhandlungen* 229:393–420.

Gale, A. S. 2006. The Cretaceous-Palaeogene boundary on the Brazos River, Texas: evidence for impact-generated sedimentation? *Proceedings of the Geologists' Association* 117:173–85.

Gallagher, W. B., C. E. Campbell, J. W. M. Jagt, and E. W. A. Mulder. 2005. Mosasaur (Reptilia, Squamata) material from the Cretaceous-Tertiary boundary interval in Missouri. *Journal of Vertebrate Paleontology* 25:473–75.

Geological Survey of Canada. 1965a. Geology North Saskatchewan River Saskatchewan-Alberta. Map 1163A. 1:1,000,000. Surveys and Mapping Division, Canada.

———. 1965b. Geology South Saskatchewan River Saskatchewan-Alberta. Map 1165A. 1:1,000,000. Surveys and Mapping Division, Canada.

Gheerbrant, E. 2009. Paleocene emergence of elephant relatives and the rapid radiation of African ungulates. *Proceedings of the National Academy of Sciences* 106:10717–21.

Gheerbrant, E., and H. Astibia. 1994. Un Nouveau Mammifère du Maastrichtien de Laño (Pays Basque Espagnol). *Comptes Rendus de l'académie des Sciences de Paris,* 2nd ser., 318:1125–31.

———. 1999. The Upper Cretaceous Mammals From Laño (Spanish Basque Country). *Estudios del Museo de Ciencias Naturales de Alava* 14:295–323.

Gheerbrant, E., D. P. Domning, and P. Tassy. 2005. Paenungulata (sirenia, Proboscidea, Hyracoidea, and relatives). In Rose and Archibald, *The Rise of Placental Mammals,* 84–105.

Gittenberger, E. 1991. What about non-adaptive radiation? *Biological Journal of the Linnean Society* 43:263–72.

Godinot, M., and G. V. R. Prasad. 1994. Eutherian tarsal bones from the Late Cretaceous of India. *Journal of Paleontology* 68:892–902.

Godthelp, H., M. Archer, R. L. Cifelli, S. J. Hand, and C. F.

Gilkeson. 1992. Earliest known Australian Tertiary mammal fauna. *Nature* 356:514–16.

Goldin, T. A., and H. J. Melosh. 2009. Self-shielding of thermal radiation by Chicxulub impact ejecta: Firestorm or fizzle? *Geology* 37:1135–38.

Goodman, M., K. N. Sterner, M. Islam, M. Uddin, C. C. Sherwood, P. R. Hoff, Z.-C. Hou, L. Lipovich, H. Jia, L. I. Grossman, and D. E. Wildman. 2009. Phylogenomic analyses reveal convergent patterns of adaptive evolution in elephant and human ancestries. *Proceedings of the National Academy of Sciences* 106:20824–29.

Gradstein, F., J. Ogg, and A. Smith, eds. 2004. *A Geologic Time Scale.* Cambridge: Cambridge University Press.

Gregory, W. K. 1910. The orders of mammals. *Bulletin of the American Museum of Natural History* 27:1–524.

Gregory, W. K., and G. G. Simpson. 1926. Cretaceous mammal skulls from Mongolia. *American Museum Novitates* 225:1–20.

Habib, D., and F. Saeedi. 2007. The *Manumiella seelandica* global spike: Cooling during regression at the close of the Maastrichtian. *Palaeogeography, Palaeoclimatology, Palaeoecology* 255:87–97.

Haile, J., D. G. Froese, R. D. E. MacPhee, R. G. Roberts, L. J. Arnold, A. V. Reyes, M. Rasmussen, R. Nielsen, B. W. Brook, S. Robinson, M. Demuro, M. T. P. Gilbert, K. Munch, J. J. Austin, A. Cooper, I. Barnes, P. Möller, and E. Willerslev. 2009. Ancient DNA reveals late survival of mammoth and horse in interior Alaska. *Proceedings of the National Academy of Sciences* 106:22352–57.

Haltennorth, T., and H. Diller. 1992. *Mammals of Africa including Madagascar.* London: Collins.

Hartman, J. 2002. The Hell Creek Formation and the early picking of the Cretaceous-Tertiary boundary in the Williston Basin. In Hartman, Johnson, and Nichols, *The Hell Creek Formation and the Cretaceous-Tertiary Boundary in the Northern Great Plains,* 1–7.

Hartman, J. H., K. R. Johnson, and D. J. Nichols, eds. 2002. *The Hell Creek Formation and the Cretaceous-Tertiary Boundary in the Northern Great Plains: An Integrated Continental Record of the End of the Cretaceous.* Geological Society of America Special Paper 361. Boulder, CO: Geological Society of America.

Harvey, M. C., S. C. Brassell, C. M. Belcher, and A. Montanari. 2008. Combustion of fossil organic matter at the K-P boundary. *Geology* 36:355–58.

Hasegawa, M., J. L. Thorne, and H. Kishino. 2003. Time scale of eutherian evolution estimated without assuming a constant rate of molecular evolution. *Genes and Genetic Systems* 78:267–83.

Hildebrand, A. R., G. T. Penfield, D. A. Kring, M. Pilkington, A. C. Zanoguera, S. B. Jacobsen, and W. V. Boynton, W. V. 1991. Chicxulub Crater: A possible Cretaceous/Tertiary boundary impact crater on the Yucatan Peninsula, Mexico. *Geology* 19:867–71.

Hoad, T. F., ed. 1996. *The Concise Oxford Dictionary of English Etymology.* Oxford: Oxford University Press.

Hoganson, J. W., and E. C. Murphy. 2002. Marine Breien Member (Maastrichtian) of the Hell Creek Formation in North Dakota: Stratigraphy, vertebrate fossil record, and age. In Hartman, Johnson, and Nichols, *The Hell Creek Formation and the Cretaceous-Tertiary Boundary in the Northern Great Plains,* 247–69.

Holroyd, P. A., and J. C. Mussell. 2005. Macroscelidea and Tubulidentata. In Rose and Archibald, *The Rise of Placental Mammals,* 71–83.

Holroyd, P. A., G. P. Wilson, and J. H. Hutchison. 2009. Turtle Diversity through the Latest Cretaceous of the Hell Creek Formation, Montana. *North American Paleontological Convention (NAPC 2009): Abstracts,* 146, www.napc2009.org/technical-program-and-abstracts.

Hooker, J. J. 2005. Perissodactyla. In Rose and Archibald, *The Rise of Placental Mammals,* 199–214.

Hornaday, W. T. 1886. The extermination of the American bison. *Report of the National Museum, 1886–'87:*369–548.

Horner, J. R. 2009. Hell Creek Formation dinosaur census reveals abundant *Tyrannosaurus. North American Paleontological Convention (NAPC 2009): Abstracts,* 149–50, www.napc2009.org/technical-program-and-abstracts.

Horner, J. R., and M. B. Goodwin. 2009. Extreme cranial ontogeny in the Upper Cretaceous dinosaur *Pachycephalosaurus. PLoS ONE* 4:e7626, http://doi:10.1371/journal.pone.0007626.

Hu, Y., J. Meng, Y. Yuanqing Wang, and C. Chuankui Li. 2005. Large Mesozoic mammals fed on young dinosaurs. *Nature* 433:149–52.

Hug, L. A., and A. J. Roger. 2007. The impact of fossils and taxon sampling on ancient molecular dating analyses. *Molecular Biology and Evolution* 24:1889–97.

Hunter, J. P., and J. D. Archibald. 2002. Mammals from the end of the Age of Dinosaurs in North Dakota and southeastern Montana. In Hartman, Johnson, and Nichols, *The Hell Creek Formation and the Cretaceous-Tertiary Boundary in the Northern Great Plains,* 191–215.

Hunter, J. P., and C. M. Janis. 2006a. Spiny Norman in the Garden of Eden? Dispersal and early biogeography of Placentalia. *Journal of Mammalian Evolution* 13:89–123.

———. 2006b. "Garden of Eden" or "Fool's Paradise"? Phylogeny, dispersal, and the southern continent hypothesis of placental mammal origins *Paleobiology* 32:339–44.

Hunter, J. P., and D. A. Pearson. 1996. First record of Lancian (Late Cretaceous) Mammals from the Hell Creek Formation of southwestern North Dakota, USA. *Cretaceous Research* 17:633–64.

Hurlbert, S., and J. D. Archibald. 1995. No evidence of sudden (or gradual) dinosaur extinction at the K/T boundary. *Geology* 23:881–84.

Hutchison, J. H. 1982. Turtle, crocodilian, and champsosaur diversity changes in the Cenozoic of the north-central region of western United States. *Palaeogeography, Palaeoclimatology, Palaeoecology* 37:149–64.

Hutchison, J. H., P. A. Holroyd, P. Gregory, and G. P. Wilson. 2004. The possible role of climate on successive turtle assemblages from the Upper Cretaceous Hell Creek Formation. *Journal of Vertebrate Paleontology* 24 (Supplement to 3):73A.

Iglesias, A., P. Wilf, K. R. Johnson, A. B. Zamuner, N. R. Cúneo, S. D. Matheos, and B. D. Singer. 2007. A Paleocene lowland macroflora from Patagonia reveals significantly greater richness than North American analogs. *Geology* 35:947–50.

Jackson, P. C. 1981. Geological Highway Map of Alberta. 2nd ed. 1 inch to 4 miles. The Canadian Society of Petroleum Geologists. Geological Highway Map Series.

Ji, Q., Z.-X. Luo, C.-X. Yuan, J. R. Wible, J.-P. Zhang, and J. A. Georgi. 2002. The earliest known eutherian mammal. *Nature* 416:816–22.

Johnson, K. R. 1992. Leaf-fossil evidence for extensive floral extinction at the Cretaceous-Tertiary boundary, North Dakota. *Cretaceous Research* 13:91–117.

Johnson, P. A., and R. C. Fox. 1984. Paleocene and Late Cretaceous mammals from Saskatchewan, Canada. *Palaeontographica, Abteilung A* 186:163–222.

Jones, M. 2008. Who was more important: Lincoln or Darwin? *Newsweek*, June 28.

Keller, G., T. Adatte, S. Gardin, A. Bartolini, and S. Bajpai. 2008. Main Deccan volcanism phase ends near the K-T boundary: Evidence from the Krishna-Godavari Basin, SE India. *Earth and Planetary Science Letters* 268:293–311.

Keller, G., T. Adatte, W. Stinnesbeck, M. Rebolledo-Vieyra, J. Urrutia-Fucugauchi, U. Kramar, and D. Stüben. 2004. Chicxulub impact predates the K-T boundary mass extinction. *Proceedings of the National Academy of Sciences* 101:3753–58.

Kennett, J. P., R. E. Houtz, P. B. Andrews, A. R. Edwards, V. A. Gostin, M. Hajos, M. A. Hampton, D. G. Jenkins, S. V. Margolis, A. T. Ovenshine, and K. Perch-Nielsen. 1974. Development of the Circum-Antarctic Current. *Science* 186:144–47.

Kielan-Jaworowska, Z. 1978. Evolution of the therian mammals in the Late Cretaceous of Asia. *Palaeontologia Polonica* 38:5–41.

Kielan-Jaworowska, Z., R. L. Cifelli, and Z-X. Luo. 2004. *Mammals from the Age of Dinosaurs: Origins, Evolution, and Structure*. New York: Columbia University Press.

Kingdon, J. 1982. *East African Mammals Volume IIID Bovids*. Chicago: University of Chicago Press.

Kirchner, J. W., and A. Weil. 2000. Delayed biological recovery from extinctions throughout the fossil record. *Nature* 404:177–80.

Koenigswald, W. von, and H.-P. Schierning. 1987. The ecological niche of an extinct group of mammals, the early Tertiary apatemyids. *Nature* 326:595–97.

Kozak, K. H., D. W. Weisrock, and A. Larson. 2006. Rapid lineage accumulation in a non-adaptive radiation: Phylogenetic analysis of diversification rates in eastern North American woodland salamanders (Plethodontidae: *Plethodon*). *Proceedings of the Royal Society B* 273:539–46.

Krause, D. W. 1986. Competitive exclusion and taxonomic displacement in the fossil record: The case of rodents and multituberculates in North America. *University of Wyoming Contributions to Geology Special Paper* 3:95–117.

———. 2001. Fossil molar from a Madagascar marsupial. *Nature* 412:497–98.

Kuiper, K. F., A. F. J. Deino, F. J. Hilgen, W. Krijgsman, P. R. Renne, and J. Wijbran. R. 2008. Synchronizing rock clocks of earth history. *Science* 320:500–504.

Kusuhashi, N., N. Ikegami, and H. Matsuoka. 2008. Additional mammalian fossils from the Upper Cretaceous Mifune Group, Kumamoto, western Japan. *Paleontological Research* 12:199–203.

Larson, G., and S. Martin. 2003. *The Complete Far Side, 1980–1994*. 2 vols. Kansas City, MO: Andrew McMeel Publishing.

Lemke, R. W. 1960. Geology of the Souris River Area North Dakota. *Geological Survey Professional Paper* 325:1–133.

LePage, D. 2004. *Avibase: The World Bird Database*. Port Rowan, ON: Bird Studies Canada, www.bsc-eoc.org/avibase/avibase.jsp.

Lewis, J. S., G. H. Watkins, H. Hartman, and R. Prinn. 1982. Chemical consequences of major impact events on Earth. In L. T. Silver and P. H. Schultz, eds., *Geological Implications of Impacts of Large Asteroids and Comets on the Earth*, Geological Society of America Special Paper 190, 215–21. Boulder, CO: Geological Society of America.

Lillegraven, J. A. 1969. Latest Cretaceous mammals of upper part of Edmonton Formation of Alberta, Canada, and review of marsupial-placental dichotomy in mammalian evolution. *University of Kansas Paleontological Contribution*, Art. 50, Vert. 12:1–122.

Lillegraven, J. A., and J. J. Eberle. 1999. Vertebrate faunal changes through Lancian and Puercan time in Southern Wyoming. *Journal of Paleontology* 73:691–710.

Lillegraven, J. A., Z. Kielan-Jaworowska, and W. A. Clemens, eds. 1979. *Mesozoic Mammals: The First Two-thirds of Mammalian Evolution*. Berkeley: University of California Press.

Lillegraven, J. A., and M. C. McKenna. 1986. Fossil mammals from the "Mesaverde" Formation (Late Cretaceous, Judithian) of the Bighorn and Wind River basins, Wyoming, with definitions of the Late Cretaceous North American Land-Mammal "Ages." *American Museum Novitates* 2840:1–68.

Lloyd, G. T., K. E. Davis, D. Pisani, J. E. Tarver, M. Ruta, M. Sakamoto, D. W. E. Hone, R. Jennings, and M. J. Benton. 2008. Dinosaurs and the Cretaceous terrestrial revolution. *Proceedings of the Royal Society B* 275:2483–90.

Lofgren, D. L. 1995. The Bug Creek problem and the Cretaceous-Tertiary transition at McGuire Creek, Montana. *University of California Publications in the Geological Sciences* 140:1–185.

Longrich, N. R., and P. J. Currie. 2009. A microraptorine (Dinosauria-Dromaeosauridae) from the Late Cretaceous of North America. *Proceedings of the National Academy of Sciences Early Edition*, 1–6.

Luo, Z.-X. 2007. Transformation and diversification in early mammal evolution. *Nature* 450:1011–19.

Luo, Z.-X., Q. Qiang Ji, J. R. Wible, and C.-X. Yuan. 2003. An Early Cretaceous tribosphenic mammal and metatherian evolution. *Science* 302:1934–40.

Lyell, C. 1872. *Principles of Geology; or, The Modern Changes of the Earth and Its Inhabitants Considered as Illustrative of Geology*. New York: Appleton.

Mac Guckin de Slane, W., trans. 1845. *Ibn Khallikān's Biographical Dictionary*, vol. 3, pt. 1 (Wafayāt al-a 'yān wa-anbā' abnā' az-zamān, "Deaths of Eminent Men and History of the Sons of the Epoch"). Paris: Benjamin Duprat.

Martin, A. J. 2009. Dinosaur burrows in the Otway Group (Albian) of Victoria, Australia, and their relation to Cretaceous polar environments. *Cretaceous Research* 30:1223–37.

Matsumato, R., and S. E. Evans. 2009. Choristoderes and the freshwater assemblages of Laurasia. In Á. D. Buscalioni and M. F. Martínez, eds., *Abstracts of the 10th International Meeting of Mesozoic Terrestrial Ecosystems and Biota, Teruel*, 69–71. Madrid: UA Ediciones.

Mayor, A. 2000. *The First Fossil hunters: Paleontology in Greek and Roman Times*. Princeton: Princeton University Press.

McDowell, S. B., Jr. 1958. The Greater Antillean insectivores. *Bulletin of the American Museum of Natural History* 115:113–214.

McGown, C. 2001. *The Dragon Seekers*. Cambridge: Perseus Publishing.

McKenna, M. C. 1975. Toward a phylogenetic classification of the Mammalia. In W. P. Luckett and F. S. Szalay, eds., *Phylogeny of the Primates: A Multdisciplinary Approach*, 21–46. New York: Plenum.

McKenna, M. C., and S. Bell. 1997. *Classification of Mammals: Above the Species Level*. New York: Columbia University Press.

McLean, D. M. 1995. K-T transition greenhouse and embryogenesis dysfunction in the dinosaurian extinctions. *Journal of Geological Education* 43:517–28.

Meng, J., Y. Hu, and C. K. Li. 2003. The osteology of *Rhombomylus* (Mammalia, Glires): Implications for phylogeny and evolution of Glires. *Bulletin of the American Museum of Natural History* 275:1–247.

Meng, J., and A. R. Wyss. 2001. The morphology of *Tribosphenomys* (Rodentiaformes, Mammalia): Phylogenetic implications for basal Glires. *Journal of Mammalian Evolution* 8:1–71.

———. 2005. Glires (Lagomorpha, Rodentia). In Rose and Archibald, *The Rise of Placental Mammals*, 145–58.

Meredith, R. W., M. Westerman, J. A. Case, and M. S. Springer. 2008. A phylogeny and timescale for marsupial evolution based on sequences for five nuclear genes. *Journal of Mammalian Evolution* 15:1–36.

Montellano, M. 1992. Mammalian fauna of the Judith River Formation (Late Cretaceous, Judithian), northcentral Montana. *University of California Publications in Geological Sciences* 136:1–115.

Morgan, J. V. 2008. Comment on "Determining chondritic impactor size from the marine osmium isotope record." *Science* 321:1158a.

Morgan, J. V., C. Lana, A. Kersley, B. Coles, C. Belcher, S. Montanari, E. Diaz-Martinez, A. Barbosa, V. Neumann. 2006. Analyses of shocked quartz at the global K-P boundary indicate an origin from a single, high-angle, oblique impact at Chicxulub. *Earth and Planetary Science Letters* 251:264–79.

Muizon, C. de, and C. Argot. 2003. Comparative anatomy of the didelphimorphs marsupials from the early Palaeocene of Bolivia (*Pucadelphys, Andinodelphys,* and *Mayulestes*). Palaeobiologic implications. In M. Jones, C. Dickman, and M. Archer, eds., *Predators with Pouches: The Biology of Carnivorous Marsupials,* 42–63. Sydney: Surrey Beatty & Sons.

Muizon, C. de, and R. L. Cifelli. 2000. The "condylarths" (archaic Ungulata, Mammalia) from the early Palaeocene of Tiupampa (Bolivia): Implications on the origin of the South American ungulates. *Geodiversitas* 22:47–150.

Murphy, W. J., E. Eizirik, W. E. Johnson, Y. P. Zhang, O. A. Ryder, and S. J. O'Brien. 2001. Molecular phylogenetics and the origins of placental mammals. *Nature* 409:614–18.

Murphy, W. J., E. Eizirik, S. J. O'Brien, O. Madsen, M. Scally, C. J. Douady, E. Teeling, O. A. Ryder, M. J. Stanhope, W. W. de Jong, and M. S. Springer. 2003. Resolution of the early placental mammal radiation using Bayesian phylogenetics. *Science* 294:2348–51.

Murphy, W. J., D. M. Larkin, A. E. der Wind, and 22 others. 2005. Dynamics of mammalian chromosome evolution inferred from multispecies comparative maps. *Science* 309:613–17.

Nessov (Nesov), L.A. 1993. New Mesozoic mammals of middle Asia and Kazakhstan and comments about evolution of theriofaunas of Cretaceous coastal plains of Asia [in Russian with English summary]. *Trudy Zoologiceskogo Instituta, Rossijskoj Akademii Nauk* 249:105–33.

———. 1997. *Cretaceous Non-marine Vertebrates of Northern Eurasia* [in Russian, with English summary]. University of Saint Petersburg, Institute of the Earth Crust, Saint Petersburg. 218 pp. Posthumous edition by L. B. Golovneva and A. O. Averianov.

Nessov (Nesov), L. A., J. D. Archibald, and Z. Kielan-Jaworowska. 1998. Ungulate-like mammals from the Late Cretaceous of Uzbekistan and a phylogenetic analysis of Ungulatomorpha. *Bulletin of the Carnegie Museum of Natural History* 34:40–88.

Newell, N. D. 1967. Revolutions in the history of life. *Geological Society of America Special Paper* 89:63–91.

———. 1987. Paleobiology's golden age. *Palaios* 2:305–9.

Nichols, D. J., and K. R. Johnson. 2008. *Plants and the K-T Boundary.* Cambridge: Cambridge University Press.

Nishihara, H., S. Maruyamab, and N. Okada. 2009. Retroposon

analysis and recent geological data suggest near-simultaneous divergence of the three superorders of mammals *Proceedings of the National Academy of Sciences* 106:5235–40.

Novacek, M. J., G. W. Rougier, J. R. Wible, M. C. McKenna, D. Dashzeveg, and I Horovitz. 1997. Epipubic bones in eutherian mammals from the Late Cretaceous of Mongolia. *Nature* 389:483–86.

O'Leary, M. A. 2010. An anatomical and phylogenetic study of the osteology of the petrosal of extant and extinct artiodactylans (Mammalia) and relatives. *Bulletin of the American Museum of Natural History* 335:1–206.

Omland, K. E. 1997. Correlated rates of molecular and morphological evolution. *Evolution* 51:1381–93.

Oms, O., J. Dinarès-Turell, E. Vicens, R. Estrada, B. Vila, A. Galobart, A. M. Bravo. 2007. Integrated stratigraphy from the Vallcebre Basin (southeastern Pyrenees, Spain): New insights on the continental Cretaceous-Tertiary transition in southwest Europe. *Palaeogeography, Palaeoclimatology, Palaeoecology* 255:35–47.

Owen R., 1841. Observations on the fossils representing the *Thylacotherium Prevostii* Valenciennes, with reference to the doubts of its mammalian and marsupial nature recently promulgated: And on the *Phascolvtherium Bucklandi. Transactions of the Geological Society of London* (2) 6:47–65.

———. 1842. Report on British fossil reptiles, pt. 2. *Report of the Eleventh Meeting of the British Association for the Advancement of Science held at Plymouth,* July 1841, 66–204.

———. 1871. Monograph of the fossil Mammalia of the Mesozoic formations. *Monographs of the Paleontographical Society* 33:1–115.

Padian, K. 2004. Book Review. *Priscum* 13:12–14.

———. 2004. Basal Avialae. In Weishampel, Dodson, and Osmólska, 210–31.

Paquay, F. S., G. E. Ravizza, T. K. Dalai, and B. Peucker-Ehrenbrink. 2008a. Determining chondritic impactor size from the marine osmium isotope record. *Science* 320:214–18.

———. 2008b. Response to comment on "Determining chondritic impactor size from the marine osmium isotope record." *Science* 321:1158b.

Pascual, R., M. Archer, E. O. Jaureguizar, J. L. Prado, H. Godthelp, and S. J. Hand. 1992. First discovery of monotremes in South America. *Nature* 356:704–6.

Patton, J. L., M. N. F. Da Silva, and J. R. Malcolm. 2000. Mammals of the Rio Juruá and the evolutionary and ecological diversification of Amazonia. *Bulletin of the American Museum of Natural History* 244:1–306.

Pearson, D. A., T. Schaefer, K. P. Johnson, D. J. Nichols. 2001. Palynologically calibrated vertebrate record from North Dakota consistent with abrupt dinosaur extinction at the Cretaceous-Tertiary boundary. *Geology* 29:39–42.

Pearson, D. A., T. Schaefer, K. R. Johnson, D. J. Nichols, and J. P. Hunter. 2002. Vertebrate biostratigraphy of the Hell Creek Formation in southwestern North Dakota and northwestern South Dakota. In Hartman, Johnson, and Nichols, *The Hell Creek Formation and the Cretaceous-Tertiary Boundary in the Northern Great Plains,* 145–68.

Pennisi, E. 2008. Comprehensive conservation database details threats to mammals. *Science* 322:178–79.

Pereira, T. V., F. M. Salzano, A. Mostowska, W. H. Trzeciak, A. Ruiz-Linares, J. A. B. Chies, C. Saavedra, C. Nagamachim, A. M. Hurtado, K. Hill, D. Castro-de-Guerra, W. A. Silva-Júnior, and M.-C. Bortolini. 2006. Natural selection and molecular evolution in primate PAX9 gene, a major determinant of tooth development. *Proceedings of the National Academy of Sciences* 103:5676–81.

Peters, S. E. 2008. Environmental determinants of extinction selectivity in the fossil record. *Nature* 454:626–29.

Poag, C. W., J. B. Plescia, and P. C. Molzer. 2002. Ancient impact structures on modern continental shelves: The Chesapeake Bay, Montagnais, and Toms Canyon craters, Atlantic margin of North America. In R. Gersonde, A. Deutsch, B. A. Ivanov, and F. Kyte, eds., *Deep-Sea Research, Part II,* 1081–1102. New York: Pergamon.

Pollman, C. D., and D. E. Canfield Jr. 1991. Florida. In D. F. Charles, ed., *Acidic Deposition and Aquatic Ecosystems Regional Case Studies,* 367–416. New York: Springer Verlag.

Pope, K. O. 2002. Impact dust not cause of the Cretaceous-Tertiary mass extinction. *Geology* 30:99–102.

Prasad, G. V. R., J.-J. Jaeger, A. Sahni, E. Gheerbrant, and C. K. Khajuria. 1994. Eutherian mammals from the Upper Cretaceous (Maastrichtian) inter-rappean beds of Naskal, Andhra Pradesh, India. *Journal of Vertebrate Paleontology* 14:260–77.

Prasad, G. V. R., and A. Sahni. 1988. First Cretaceous mammal from India. *Nature* 332:638–40.

Prasad, G. V. R., O. Verma, A. Sahni, V. Parmar, and A. Khosla. 2007. A Cretaceous hoofed mammal from India. *Science* 318:937.

Preston, D. J. 1993. *Dinosaurs in the Attic: An Excursion Into the American Museum of Natural History.* New York: St. Martin's Press.

Prinn, R. G., and B. Fegley Jr. 1987. Bolide impacts, acid rain, and biospheric traumas at the Cretaceous-Tertiary boundary. *Earth and Planetary Science Letters* 83:1–15.

Rasband, W. 2008. ImageJ: Image Processing and Analysis in Java. Version 1.40g. National Institutes of Health, http://rsb.info.nih.gov/ij/index.html.

Raup, D. M. 1991. *Extinction: Bad Genes or Bad Luck?* New York: Norton.

Raup, D. M., and J. J. Sepkoski Jr. 1982. Mass extinction in the marine fossil record. *Science* 215:1501–3.

Retallack, G. J. 1993. Evidence from paleosols for acid overdose at the end of the Cretaceous in Montana. *Journal of Vertebrate Paleontology* 13 (supplement to no. 3):54A.

———. 1994. A pedotype approach to latest Cretaceous and earliest Tertiary paleosols in eastern Montana. *Geological Society of America Bulletin* 106:1377–97.

———. 2004. End-Cretaceous acid rain as a selective extinction mechanism between birds and dinosaurs. In P. J. Currie, E. B. Koppelhus, M. A. Shugar, and J. L. Wright, eds., *Feathered Dragons: Studies on the Transition from Dinosaurs to Birds,* 35–64. Bloomington: Indiana University Press.

Rich, T. H., P. Vickers-Rich, T. F. Flannery, B. P. Kear, D. J. Cantrill, P. Komarower, L. Kool, D. Pickering, P. Trusler, S. Morton, N. Van Klaveren, and E. M. G. Fitzgerald. 2009. An Australian multituberculate and its palaeobiogeographic implications. *Acta Palaeontologica Polonica* 54:1–6.

Rigby, J. K., and D. L. Wolberg. 1987. The therian mammal fauna (Campanian) of Quarry 1, Fossil Forest Study Area, San Juan Basin, New Mexico. *Geological Society of America Special Paper* 209:472–93.

Robertson, D. S., M. C. McKenna, O. B. Toon, S. Hope, and J. A. Lillegraven. 2004a. Survival in the first hours of the Cenozoic. *Geological Society of America Bulletin* 116:760–68.

———. 2004b. Fireball passes and nothing burns: Comment. *Geology* 32:e50, http://doi:10.1130/0091-7613-32.1.e50.

Rogers, R. R. 1998. Sequence analysis of the Upper Cretaceous Two Medicine and Judith River Formations, Montana: Nonmarine response to the Claggett and Bearpaw marine cycles. *Journal of Sedimentary Research* 68:615–31.

Romer, A. S., and T. S. Parsons. 1977. *The Vertebrate Body.* Philadelphia: W. B. Saunders Co.

Rose, K. D., and J. D. Archibald, eds. 2005. *The Rise of Placental Mammals.* Baltimore: Johns Hopkins University Press.

Rose, K. D., R. J. Emry, T. J. Gaudin, and G. Storch. 2005. Xenarthra and Pholidota. In Rose and Archibald, *The Rise of Placental Mammals,* 106–26.

Rougier, G. W., S. Isaji, and M. Manabe. 2007. An Early Cretaceous mammal from the Kuwajima Formation (Tetori Group), Japan, and a reassessment of triconodont phylogeny. *Annals of Carnegie Museum* 76:73–115.

Rowe, T. 1988. Definition, diagnosis and origin of Mammalia. *Journal of Vertebrate Paleontology* 8:241–64.

———. 1999. At the roots of the mammalian family tree. *Nature* 398:283–84.

Rowe, T., R. L. Cifelli, T. M. Lehman, and A. Weil. 1992. The Campanian age Terlingua local fauna, with a summary of the other vertebrates from the Aguja Formation, Trans-Pecos Texas. *Journal of Vertebrate Paleontology* 12:472–93.

Rowe, T., T. H. Rich, P. Vickers-Rich, M. Springer, and M. O. Woodburne. 2008. The oldest platypus and its bearing on divergence timing on platypus and echidna clades. *Proceedings of the National Academy of Sciences* 105:1238–42.

Rudwick, M. J. S. 1997. *Georges Cuvier, Fossil Bones, and Geological Catastrophes: New Translations and Interpretations of the Primary Texts.* Chicago: University of Chicago Press.

Rundell, R. J., and T. D. Price. 2009. Adaptive radiation, nonadaptive radiation, ecological speciation and nonecological speciation. *Trends in Ecology and Evolution* 24:394–99.

Ryan, M. J., and D. C. Evans. 2005. Ornithischian dinosaurs. In Currie and Koppelhus, *Dinosaur Provincial Park,* 312–48.

Salvador, A. 2006. Geologic Note: The Tertiary and the Quaternary are here to stay. *American Association of Petroleum Geologists Bulletin* 90:21–30.

Sánchez-Villagra, M. R., and K. K. Smith. 1997. Diversity and evolution of the marsupial mandibular angular process. *Journal of Mammalian Evolution* 24:119–44.

Sanderson, M. J. 1998. Reappraising adaptive radiation. *American Journal of Botany* 85:1650–55.

Sankey, J. 2009. High abundance of sharks in uppermost Hell Creek Formation, North Dakota: Sea level rise due to the Late Maastrichtian greenhouse event? *North American Paleontological Convention (NAPC 2009): Abstracts,* p. 144. www.napc2009.org/technical-program-and-abstracts.

Sargis, E. J., S. G. B. Chester, F. S. Szalay, J. D. Archibald, and A. O. Averianov. 2009. Mammalian distal humeri from the Late Cretaceous of Uzbekistan. *Acta Palaeontologca Polonica* 55:199–211.

Scannella, J. 2009. And then there was one: Synonymy consequences of *Triceratops* Cranial ontogeny. *Journal of Vertebrate Paleontology* 29 (supplement. to no. 3):177A.

Schipper, J., J. S. Chanson, F. Chiozza, and 127 others. 2008. The status of the world's land and marine mammals: Diversity, threat, and knowledge. *Science* 322:225–30.

Scholz, H., and J. H. Hartman. 2007. Fourier analysis and the extinction of unionoid bivalves near the Cretaceous-Tertiary boundary of the Western Interior, USA: Pattern, causes, and ecological significance. *Palaeogeography, Palaeoclimatology, Palaeoecology* 255:48–63.

Schulte, P., L. Alegret, I. Arenillas, and 38 others. 2010. The Chicxulub asteroid impact and mass extinction at the Cretaceous-Paleogene boundary. *Science* 327:1214–18.

Schultz, P. H., and S. D'Hondt. 1996. Cretaceous-Tertiary (Chicxulub) impact angle and its consequences. *Geology* 24:963–67.

Scott, G. R., and C. G. Turner II. 1997. *The Anthropology of Modern Human Teeth: Dental Morphology and Its Variation in Recent Human Populations.* Cambridge: Cambridge University Press.

Seiffert, E. R., E. L. Simons, T. M. Ryan, T. M. Bown, and Y. Attia. 2007. New remains of Eocene and Oligocene Afrosoricida (Afrotheria) from Egypt, with implications for the origin(s) of afrosoricid zalambdodonty. *Journal of Vertebrate Paleontology* 27:963–72.

Setoguchi, T., T. Tsubamoto, H. Hanamura, and K. Hachiya. 1999. An early Late Cretaceous mammal from Japan, with reconsideration of the evolution of tribosphenic molars. *Paleontological Research* 3:18–28.

Sheehan, P. M., D. E. Fastovsky, C. Barreto, and R. G. Hoffman. 2000. Dinosaur abundance was not declining in a "3 m gap" at the top of the Hell Creek Formation, Montana and North Dakota. *Geology* 28:523–26.

Sheehan, P. M., D. E. Fastovsky, R. G. Hoffman, C. B. Berghaus, and D. L. Gabriel. 1991. Sudden extinction of the dinosaurs: Latest Cretaceous, upper Great Plains, USA. *Science* 254:835–39.

Sidlauskas, B. 2008. Continuous and arrested morphological diversification in sister clades of characiform fishes: A phylomorphospace approach. *Evolution* 62: 3135–56.

Signor, P. W., III, and J. H. Lipps. 1982. Sampling bias, gradual extinction patterns, and catastrophes in the fossil record. In L. T. Silver and P. H. Schultz, eds., *Geological Implications of Impacts of Large Asteroids and Comets on the Earth.* Geological Society of America Special Paper 190, 291–96. Boulder, CO: Geological Society of America.

Silcox, M. T., J. I. Bloch, E. J. Sargis, and D. M. Boyer. 2005. Euarchonta (Dermpotera, Scadentia, Primates). In Rose and Archibald, *The Rise of Placental Mammals*, 127–44.

Simmons, N. B., K. L. Seymour, J. Habersetzer, and G. F. Gunnell. 2008. Primitive Early Eocene bat from Wyoming and the evolution of flight and echolocation. *Nature* 451:818–22.

Simpson, G. G. 1928a. Affinities of the Mongolian Cretaceous insectivores. *American Museum Novitates* 330:1–11.

———. 1928b. *A Catalogue of the Mesozoic Mammalia in the Geological Department of the British Museum.* London: Trustees of the British Museum, London.

———. 1929. American Mesozoic Mammalia. *Memoir of the Peabody Museum Yale University* 3:1–235.

———. 1945. The principles of classification and a classification of mammals. *Bulletin of the American Museum Natural History* 85:1–350.

Smith, A. L., D. G. Smith, and B. M. Funnell. 1994. *Atlas of Mesozoic and Cenozoic Coastlines.* Cambridge: Cambridge University Press.

Spaulding, M., M. A. O'Leary, and J. Gatesy. 2009. Relationships of Cetacea (Artiodactyla) among mammals increased taxon sampling alters interpretations of key fossils and character evolution. *PLoS One* 4:1–14.

Springer, M S., W. J. Murphy, E. Eizirik, and S. J. O'Brien. 2003. Placental mammal diversification and the Cretaceous-Tertiary boundary. *Proceedings of the National Academy of Sciences* 100:1056–61.

Stanhope, M. J., V. G. Waddell, O. Madsen, W. W. de Jong, S. B. Hedges, G. C. Cleven, D. Kao, and M. S. Springer. 1998. Molecular evidence for multiple origins of Insectivora and for a new order of endemic African insectivore mammals. *Proceedings of the National Academy of Sciences* 95:9967–72.

Storer, J. E. 1991. The mammals of the Gryde local fauna, Frenchmen Formation (Maastrichtian: Lancian), Saskatchewan. *Journal of Vertebrate Paleontology* 11:350–69.

Strahan, R., ed. 2000. *The Mammals of Australia.* 2nd ed. Sydney: Reed New Holland.

Swindler, D. R. 2002. *Primate Dentition: An Introduction to the Teeth of Non-human Primates.* Cambridge: Cambridge University Press.

Szalay, F. S., and E. J. Sargis. 2006. Cretaceous therian tarsals and the metatherian-eutherian dichotomy. *Journal of Mammalian Evolution* 13:171–210.

Theodor, J. M., K. D. Rose, and J. Erfurt. 2005. Artiodacyla. In Rose and Archibald, *The Rise of Placental Mammals*, 215–33.

Thompson, A. 2008. Dinosaur Killer May Have Been Volcanism, Not Asteroid. LiveScience, www.livescience.com/animals/081215-agu-volcanism-dinosaurs.html.

Thompson, A., N. C. Arens, and A. H. Jahren. 2009. Vegetation indicators of environmental stress precede the Cretaceous/Tertiary boundary. *North American Paleontological Convention (NAPC 2009): Abstracts*, p. 151. www.napc2009.org/technical-program-and-abstracts.

Torrens, H. 1992. When did the dinosaur get its name? *New Scientist* 134:44–44.

Ussher, J. 1650. *Annales veteris testamenti, a prima mundi origine deducti (Annals of the Old Testament, Deduced from the First Origins of the World).* London.

Van Valen, L. 1964. A possible origin for rabbits. *Evolution* 19:484–91.

Van Valen, L., and R. E. Sloan. 1966. The extinction of the multituberculates. *Systematic Zoology* 15:261–78.

Varricchio, D. J., A. J. Martin, and Y. Katsura. 2007. First trace and body fossil evidence of a burrowing, denning dinosaur. *Proceedings of the Royal Society B* 274:1361–68.

Vullo, R., E. Gheerbrant, C. de Muizon, and D. Néraudeau. 2009. The oldest modern therian mammal from Europe and its bearing on stem marsupial paleobiogeography. *Proceedings of the National Academy of Sciences* 106:19910–15.

Waddell, P. J., Y. Cao, M. Hasegawa, D. P. Mindell. 1999. Assessing the Cretaceous superordinal divergence times within birds and placental mammals by using whole mitochondrial protein sequences and an extended statistical framework. *Systematic Biology* 48:119–37.

Waddell, P. J., N. Okada, and M. Hasegawa. 1999. Towards resolving the interordinal relationships of placental mammals. *Systematic Biology* 48:1–5.

Walker, C. A. 1981. New subclass of birds from the Cretaceous of South America. *Nature* 292:51–53.

Wang, S. C., and P. Dodson. 2006. Estimating the diversity of dinosaurs. *Proceedings of the National Academy of Sciences* 103:13601–5.

Warren, W. C., W. H. L. Deana, J. A. M. Graves, and 99 others. 2008. Genome analysis of the platypus reveals unique signatures of evolution. *Nature* 453:175–84.

Weil, A., and W. A. Clemens. 1998. Aliens in Montana: Phylogenetically and biogeographically diverse lineages contributed to an earliest Cenozoic community. *Geological Society of America Abstracts with Programs* 30:69–70.

Weishampel, D. B., P. M. Barrett, P. M. Coria, J. Le Loeuff, X. Xing, Z. Xijin, A. Sahni, E. M. P. Gomani, and C. R. Noto. 2004. Dinosaur distribution. In Weishampel, Dodson, and Osmólska, *The Dinosauria*, 517–613.

Weishampel, D. B., P. Dodson, and H. Osmólska, eds. 2004. *The Dinosauria*. 2nd ed. Berkeley: University of California Press.

Whitaker, J. O., Jr. 1980. *The Audubon Society Field Guide to North American Mammals*. New York: Alfred A. Knopf.

Wible, J. R., M. J. Novacek, and G. W. Rougier. 2004. New date on the skull and dentition in the Mongolian Late Cretaceous eutherian mammal *Zalambdalestes*. *Bulletin of the American Museum of Natural History* 281:1–144.

Wible, J. R., G. W. Rougier, M. J. Novacek, and R. J. Asher. 2007. Cretaceous eutherians and Laurasian origin for placental mammals near the K/T boundary. *Nature* 447:1003–6.

———. 2009. The eutherian mammal *Maelestes gobiensis* from the Late Cretaceous of Mongolia and the phylogeny of Cretaceous Eutheria. *Bulletin of the American Museum of Natural History* 327:1 123.

Wilf, P., and K. R. Johnson. 2004. Land plant extinction at the end of the Cretaceous: A quantitative analysis of the North Dakota megafloral record. *Paleobiology* 30:347–68.

Wilf, P., K. R. Johnson, and B. Huber. 2003. Correlated terrestrial and marine evidence for global climate changes before mass extinction at the Cretaceous-Paleogene boundary. *Proceedings of the National Academy of Sciences of the United States* 100:599–604.

Wilson, D. E., and D. M. Reeder, eds. 2005. *Mammal Species of the World*. Baltimore: Johns Hopkins University Press.

Wilson, G. P. 2005. Mammalian faunal dynamics during the last 1.8 million years of the Cretaceous in Garfield County, Montana. *Journal of Mammalian Evolution* 12:53–76.

———. 2009. High-resolution mammalian faunal dynamics leading up to and across the Cretaceous-Tertiary boundary in northeastern Montana. *North American Paleontological Convention (NAPC 2009): Abstracts*, 149–50, www.napc2009.org/technical-program-and-abstracts.

Wolfe, J. A. 1991. Palaeobotanical evidence for a June "impact winter" at the Cretaceous/Tertiary boundary. *Nature* 352:420–23.

Zhang, Y. 2009. Late Cretaceous Mammalian Fauna from the Hell Creek Formation, Southeastern Montana. Unpublished master's thesis, San Diego State University.

Zhonghe, Z., P. M. Barrett, and J. Hilton. 2003. An exceptionally preserved Lower Cretaceous ecosystem. *Nature* 421:807–11.

INDEX